Morgans to 1997

Morgans to 1997

A collector's guide
by Roger Bell

MOTOR RACING PUBLICATIONS LTD
Unit 6, The Pilton Estate, 46 Pitlake, Croydon CR0 3RY, England

First published 1997

British Library Cataloguing in Publication Data

Bell, Roger
Morgans to 1997 : a collector's guide
1. Morgan automobile – Collectors and collecting
I. Title
629.2'222

ISBN 1-899870-20-2

Printed in Great Britain by
The Amadeus Press Ltd
Huddersfield, West Yorkshire

Contents

Forward to the past

Morgan was facing a bleak future in the early Sixties. Its sportscars were slated as crude and old fashioned, its production methods as archaic and inefficient. In pandering to the old ways – to tradition, craftsmanship and nostalgia – Morgan was alienating buyers. "We cut production," recalls Peter Morgan, who has headed the family firm since the death of his father, Morgan's founder. "To make it look as though we were selling, we'd swap cars in the showroom of our London agent for different coloured ones after five or six weeks. We had unsold cars all over the place. No-one wanted the old styling."

How times have changed. The 'classic' look that threatened Morgan's existence when Beatlemania and mini skirts were sweeping the country was, by the end of the Sixties, to prove its salvation. Morgan's biggest problem by the turn of the swinging decade was not selling its cars but making enough of them to meet demand. It's been that way ever since. Even though legislation temporarily forced Morgan out of the lucrative US market, delivery dates stretched first into months, then – notoriously – into years.

Explaining the dramatic turnaround of fortunes in an interview for this book, Peter Morgan said: "Winning the 2-litre class at Le Mans helped us a lot, particularly abroad." So, ironically, did Morgan's first and only serious postwar attempt to modernize the styling of its cars. "I didn't have the guts to show the new Plus 4 Plus at Earls Court on its own," Peter Morgan recalls. And just as well. In their rejection of the sleek newcomer, buyers came to their senses and voted with their wallets for the traditional models instead. The *real* Morgans were no longer perceived as dated anachronisms, but as bespoke, coachbuilt classics of great character and appeal. Nostalgia was back in vogue.

By preserving old values and methods – you'll find no robots at Morgan's labour-intensive factory – and a style evocative of a golden motoring age, Morgan has since moved conservatively forward while rooted firmly in the past. Its compelling funsters have their faults – a dreadful ride, poor weather protection, minimal room, laughable ergonomics – but they provide the perfect antidote to modern motoring. "You need a sense of humour to run a Morgan," admits Peter Morgan.

We're not concerned here with mass-made consumer durables so much as soulmates, providers of entertainment, beguiling pieces of hand-crafted machinery that draw you into the fun of driving as surely as most refined and cosseting tin boxes distance you from it. If all you want from your car is to get from A to B as smoothly, quietly and uneventfully as possible, Morgans are not for you. Nor is this book.

Peter Morgan once told the American magazine *Car and Driver*: "I can give you more reasons for *not* buying a Morgan. You haven't got the comfort of a saloon and you can't lock them up...what you have got is style, performance, handling and a car that's more *fun* than anything else on the road." Owning a Ferrari never stopped Peter Morgan from commuting to work in a Plus 8.

While Morgan four-wheelers have certainly advanced since the Thirties – to pass stringent safety and emissions laws,

The Morgan Motor Company has always been a family affair, and these days the firm's welfare is in the capable hands of Peter Morgan, Chairman, and his son Charles, who is in the driving seat as Operations Director.

they have needed to – it is their charm that makes them irresistible to rat-race escapees. Morgan's survival, sometimes against the odds, is testimony to more than a unique product. It is also a tribute to an enlightened family management not driven by the usual business forces of expansionist ambition and greed. It may have made mistakes, but it has never rashly gambled. Peter Morgan explains the ethos: "We have tried to keep as downmarket as possible, to make a good low-priced car. It hasn't been easy, and it's getting more difficult."

The story of Morgan spans eight decades, two factories (the present one at Malvern Link, in Worcestershire, has been operating since 1918), four generations of the Morgan family and, in round figures, some 60,000 cars. Exact numbers are open to conjecture. A company information pack mentions a figure of 30,000 four-wheelers. However, totting up totals, model by model, the tally would appear to be closer to 20,000 in 1997 than 30,000. How many survive is pure guesswork, but is probably the majority.

Morgan's history embraces two overlapping eras. The first, the trike period, started in 1910, when the infant motor industry, barely a decade old in Britain, was still spawning adventurous newcomers, Morgan among them. "We were very unwelcome locally in father's time," recalls Peter Morgan. "Malvern was seen as a place for schools and retired colonels, not for industry." Attitudes have since changed. "Today, we're very popular, especially for bringing foreign visitors to Malvern's hotels." Morgan provides employment for some 130 people – many of whom it has trained for the job – and the company has an annual turnover of more than £10 million.

Extended by World War Two, the first era ran for over 40 years, until the early Fifties, when Morgan finally axed the old and unfashionable three-wheelers on which its reputation and success was founded – and HFC Morgan's fortune made. "Father made a lot of money in the

Twenties," recalls his son. "I don't think he ever had to borrow." One legacy of those healthy prewar profits – slashed (by reduced prices) with the arrival of the Austin Seven and the depression – is Morgan's solid financial footing today. "I've never wanted to be a large manufacturer, making thousands of cars a year. Five hundred is about right." Morgan finances its own capital ventures and owns the Malvern factory. "We did have an overdraft facility in the Sixties, but we never called upon it," says Peter Morgan.

The second period, still going strong after the odd hiccup, started in 1936 when Morgan launched its first commercial four-wheeler, the 4/4 – signifying four wheels and four cylinders. Although trikes are covered in the first two chapters, the focus of this book is on the second, four-wheeler period.

In 1990, when Morgan was featured in the BBC TV's *Troubleshooter* series, Sir John Harvey-Jones, former ICI chairman, gave Morgan five years – unless it raised prices significantly, modernized its cars and production methods, and reduced its notoriously long delivery times. Morgan rejected Sir John's proposals. Said Charles Morgan, grandson of the founder: "His methods would result in making many changes to the traditional way the Morgan is built. We believe our policy of gradual and carefully considered change will enable us to maintain the car's qualities and unique appeal."

Peter Morgan looks back on *Troubleshooter* with amusement, not anger. "I liked Harvey-Jones and the programme was tremendous publicity. We got masses of orders – the phone in the sales office never stopped ringing for a month." Whether all those orders were taken up several years later, Peter Morgan rather doubts. "Many came from the wrong sort of buyers. They didn't really know what they would be getting. We've never tried to pretend a Morgan is something it's not." Morgan turned down a request for a sequel to *Troubleshooter*. "It would have looked as though we were touting for publicity."

Although Morgan has dragged some aspects of its business into the Nineties (with computerized stock control, a new high-tech paintshop, catalytic converters and airbags, for instance), its cars are still rooted firmly in the past. For how much longer, though, remains to be seen. Morgan concedes that the sole aim of the factory's racing programme – Charles Morgan was preparing for the '97 BPR Global Endurance GT series when this was written – is to improve its road cars. Said Peter Morgan, bowing to the inevitable: "I do foresee the time when we will have to have an alternative front suspension with more travel...and also, of course, an independent rear end. We'd have to do the whole thing..."

Change is afoot, it seems, at the world's oldest privately owned – and most traditional – car maker. But don't hold your breath.

Acknowledgements

My thanks to the following who have given up their time to assist me with this book. From the Morgan Motor Company Ltd, Peter and Charles Morgan, Mark Aston, Bill Beck, Paul Trussler. From the dealer network agents, Rob Wells (Libra Motive), Rick Bourne (Brands Hatch Morgan) and Richard Thorne (Richard Thorne Classic Cars). From the MSCC, technical experts Allan Cameron, John Hancock, Chas Smith and Simon Dale-Smith. Thanks, too, to Chris Rowe, Michael Cotton and David Hodges. I am particularly grateful to *Miscellany* assistant editor Cliff Baker for providing many photographs from his splendid collection of Morgan pictures. To anyone I have overlooked, my apologies.

CHAPTER 1

Ancestry and heritage

The rector's son gets mobile

The Morgan story starts over a century ago, on August 11, 1881, when Harry Frederick Stanley Morgan – Harry or HFS for short – was born at Morton Jeffries Rectory, in Herefordshire, England. At the time, few people in Britain had heard of either Gottleib Daimler or Karl Benz, each working independently of the other in Germany to create the world's first real cars. In England's sleepy Herefordshire, the horse still reigned – but not for long. Young HFS had little time for equine transport.

As his birthplace suggests, Morgan was of solid Anglican stock. His father was the Reverend Henry George Morgan, rector of Stoke Lacy (like his father before him). As a prebendary, he was also a stand-in at the local cathedral to his bishop. It was not the most likely background for a budding businessman with an inventive streak.

HFS did not follow his father into the Church. Just the opposite: Morgan senior – a devoted family man who did what he thought best for his offspring – was to take an active and financial interest in his son's career and later business affairs. He was also a staunch defender of the three-wheeler, a competent artist and a prolific writer of letters to the technical press. He was quite well off, too.

After an education at Stone House, Broadstairs and Marlborough College, Morgan junior pursued his technical bent at the Crystal Palace Engineering College before being apprenticed, as an 18-year-old draughtsman (and at his father's instigation), to William Dean, a senior engineer of the Great Western Railway at Swindon. This, remember, was at a time when the initials GWR stood for all that was best in British engineering. What better grounding for a young designer? He was in good company, too: WO Bentley, no less, had cut his teeth in the steam sheds of the LNER. "GWR locomotives are faster than ours only because they burn superior Welsh coal," WO is alleged to have said.

HFS worked at a GWR drawing board for seven years, gaining valuable design experience, before opening a 'motor works' at Malvern Link in 1906, to the consternation of the anti-car locals: the scholarly spa town of Malvern was no place for noisy, newfangled mechanical contraptions, even less for the industry that Morgan was to bring to it later. He became an agent for Wolseley and Darracq, started a bus service (running 15-seater, 10hp Wolseleys to nearby Malvern Wells and Gloucester), then went into the car hire business.

Young Morgan's early driving career was punctuated by incidents. "My first motoring experience," he once told *Light Car*, "was in 1899 when I rode a Minerva motorcycle. A little later, a 3.5hp Benz ran away with me down a steep slope and cost my father £28 for repairs." Later, he was fined for exceeding the 12mph speed limit.

The first car that HFS owned, at the age of 21, was a British-made Eagle Tandem – a cumbersome, top-heavy tricycle on which the driver sat perched at the back over a single wheel, driven by an 8hp de Dion engine. The passenger lounged at the front, closest to the accident, in a wicker basket. "It was fast but not too reliable," HFS told *Light Car*. The struggling Eagle concern didn't survive, but

HFS Morgan's first three-wheeler was this Eagle Tandem, which he bought in 1901 whilst working as a draughtsman for the Great Western Railway. Five years later he had resigned, gone into partnership with a friend and opened a garage in Malvern, where his thoughts for the design of a lightweight three-wheeler matured.

at least one of its products was to influence, if not inspire, the first Morgan three-wheeler. So did the 7hp, two-cylinder Star that young Morgan bought. "It gave me good service for many years," he is quoted as saying. HFS shrugged aside local opposition to motorists and their horse-frightening, smoke-belching infernal contraptions. And he the rector's son, too.

Motivated by creative rather than business instincts, Morgan set about building his original Runabout for fun, not profit. Assistance in its design and construction came from the local college's machine shop and its noted engineering master, Mr Stephenson-Peach, grandson of George Stephenson of Rocket locomotive fame. The Runabout they designed and built in 1908-9 was neither motorcycle nor car, but something in between called a cyclecar. Its basis was a simple three-tube chassis, the main central one acting as a strong backbone that carried the driveshaft. The two outriggers doubled as exhaust pipes. Power came from a 7hp Peugeot V-twin engine mounted transversely up front, where it received the full brunt of the cooling airstream. The French motor was intended for a motorbike, HFS later revealed, but although a keen cyclist – Morgan had built and raced his own bike at college – he was never seriously bitten by the motorcycle bug.

From the front-end clutch, the drive was shafted through the central chassis tube to a bevel-gearbox under the driver's backside, then by chains – one for each forward gear – to the single rear wheel, sprung and partially located by a pair of quarter-elliptic springs anchored to the chassis. The two gears, high and low, were swapped by a simple dog-clutch system operated by an outside lever. There was no reverse, and none was legally required for three-wheelers weighing under 7cwt (355kg). HFS' Runabout, virtually bereft of bodywork, was exceptionally light. Neither the side-tiller steering nor the absence of front brakes was considered out of the ordinary at the time. Foot and hand-operated bands acting on the transmission provided the only means of slowing down.

The Runabout was notable for two design features that have been Morgan trademarks ever since: one was unique (and patented) independent sliding-pillar front suspension. Adopted by Lancia in the early Twenties for its revolutionary Lambda, it is still used by Morgan today, fundamentally unchanged. The other was light weight and compactness. Decades before Colin Chapman of Lotus was preaching the gospel of speed through lightness, HFS was putting theory into practice with his Runabout, which had a power-to-weight ratio of 90bhp per ton. Acceleration was said to be as swift as that of anything else on the road, regardless of power or price. HFS told *Light Car* later: "The Runabout was an instant success, due to its rigid frame, independent front suspension and light weight."

The Runabout, registration CJ 743, was well named. Morgan used it as local business wheels, initially as a solo without mudguards – indeed, with no bodywork to speak of at all, even less space to stow anything. Being light and nippy, though, it attracted much attention. With financial backing from Morgan senior, HFS was soon moved by

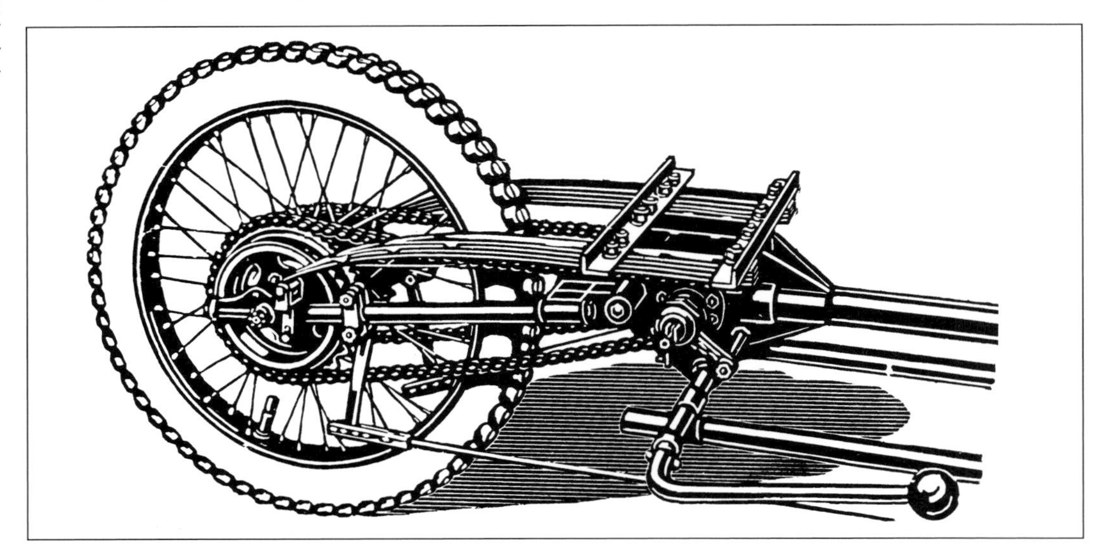

An appealing cocktail of simplicity and ingenuity. The sliding-pillar front suspension straddling a V-twin engine, and dual leaf springs flanking the single shaft and chain-driven rear wheel, were to remain familiar Morgan three-wheeler trademarks for many years.

inquiries to build production replicas at his extended garage. Thus was Morgan, car manufacturer, conceived, though its subsequent birth was not without labour pains.

Two completed single-seaters, both with stiffened-up chassis and improved driver protection, were exhibited at the motorcycle show at Olympia in 1910. HFS recalled in *Light Car* that drawings for the design patents "were done by a bright youth who is now the famous Sir John Black of Standard." The cheaper version was powered by a single-cylinder, 4hp JAP engine, the quickie by an 8hp JAP V-twin. Said a contemporary report: "It is very lightly constructed throughout...and low and rakish looking – so low, it might prove unsuitable on rough roads...the front axis has spiral springs...the rear wheel is hinged at the back of the gearbox...and is controlled by short laminated springs...splayed mudguards complete the equipment of a very cleverly designed runabout for a single rider." Note rider, not driver.

Despite such eulogies, neither model attracted a flood of orders for the fledgling Morgan operation, little known outside Malvern. Besides, the Runabouts were unproven in competition. "They were too novel," HFC told *Light Car*. "I found that demand for a two-seater would be much greater." As trials success counted for a lot in those days, Morgan persevered, with backing from an unlikely source – Richard Burbridge of Harrods, the famous Knightsbridge store in London, which became a Morgan agency. In 1996, Rover boasted in a press release that its MG F was the first car to feature in a Harrods window display. Not so. Morgan beat MG to it by 85 years.

To prove that his three-wheeled cyclecar was a cut above all the other frail contraptions taking to the roads from backstreet workshops throughout the country, Morgan turned to trialling – an early form of rallying, modelled on horse trials. In the first London-to-Exeter, he won a gold medal, the highest award. The following year he gained another 'gold' in the London-to-Edinburgh trial – no mean feat on atrocious roads with crash-damaged front suspension and a back wheel that was almost bound to hit the potholes avoided by the front ones.

When Morgan exhibited at Olympia for the second time, in 1911, he had more than competition accolades to back his nifty Runabout: his cyclecar now also had two seats, a proper steering wheel and a price tag of 85 guineas – about £90. What's more, it had been well reviewed by WG McMinnies in *Motor Cycling*. Morgan had arrived.

CHAPTER 2

The three-wheelers

Runabouts from 1912 to 1951

Several important events marked 1912 as a special year in the life of Harry Morgan. He married Hilda Ruth Day, daughter of a local vicar (she was to bear him six children – five daughters and one son, Peter); he broke the one-hour 1,100cc record at Brooklands, averaging almost 60mph (at the time, the land speed record was barely twice as fast); a Morgan won the first-ever cyclecar race held at Brooklands, driven by Harry Martin; and the Morgan Motor Company was formed as a private limited company. The ebullient Reverend HG Morgan was its chairman and backer, HFS – the firm's founder and innovator – its managing director. It was also in 1912 that Morgan made its first profit, said to be of just over £1,300.

Following the 1912 Olympia show, Morgan was inundated with more orders than he could fulfil, so several other manufacturers were approached with a view to contracting out assembly. "They fortunately turned the offer down," HFS later told *Light Car*. One can only speculate on the outcome had they not done so. "With the aid of deposits on orders, I bought machine tools, built some new workshops and, giving up my garage business, did my best to satisfy the demand." A £3,000 investment from George Morgan, whose forebears had been London merchants, also helped to finance expansion.

Morgan's early commercial success with its three-wheeled lightweights was founded on no-nonsense value-for-money design – and competition success. "I entered many hillclimbs, trials and races. Helped by owners and agents, the Runabout won plenty of awards," HFS later recalled. Morgan's low-slung cyclecars, JAP-powered to begin with, were small but quite sophisticated. They were a cheap and effective means of transport for the impecunious. Good economy and a favourable £4 road tax levied on three-wheelers – four-wheelers paid £1.00 for every RAC-rated 'horsepower' – encouraged sales. Unlike some over-ambitious rivals, Morgan resisted the temptation to design and build its own engines. There was no need, with so many proprietary units to choose from, among them the Swiss-made MAG which, with four-valve cylinder heads, gave Morgan's trike a top speed of over 70mph in 1914.

As Morgan's advertising blurb said: "It (the Runabout) is designed to meet the needs of those who require something cheaper and simpler than a car, but more comfortable than a motorcycle. To this end, particular attention has been paid to the simplicity of construction, easy accessibility of all parts, lightness and strength. The price is fixed to suit the man of moderate means, but quality has not been sacrificed to cheapness. Every part is standardised and, as an illustrated parts list is given with every machine, renewals can be made cheaply and without delay."

A Morgan advertising slogan of the time – low costs, high performance – was apt and honest. In a 1985 interview for *Car and Driver*, Peter Morgan told Charles Fox: "Father was able to make things that really worked terribly cheaply. I've never been able to make things more cheaply since he died. Besides that, he was an awfully clever business manager; he played his cards a bit carefully."

Morgan's claims were backed by competition success that

made its thumping V-twin three-wheelers the cyclecars to beat in trials events, best described as long-distance cross-country obstacle courses where gradients, mud and stones were natural hazards. Morgans excelled in hillclimbing and racing, too. The company's reputation rocketed when Gordon McMinnies of *The Cyclecar* won the 1913 French Cyclecar Grand Prix at Amiens in a carefully-prepared machine powered by a water-cooled JAP V-twin with overhead valvegear. Said HFS of the victory later in *Light Car*: "It was a wonderful performance, as he had to change a front inner tube during the event. After the race, we called this model the Grand Prix. It was popular for many years."

McMinnies' subsequent 'there-I-was' report in his magazine carried more weight than any advertising copy could have done. Orders escalated, particularly from France (Morgans were later made there under licence by the French concessionaires Darmont and Baudelocque).

Although Morgan's three-wheelers changed little conceptually in 40 years, there were many model variations

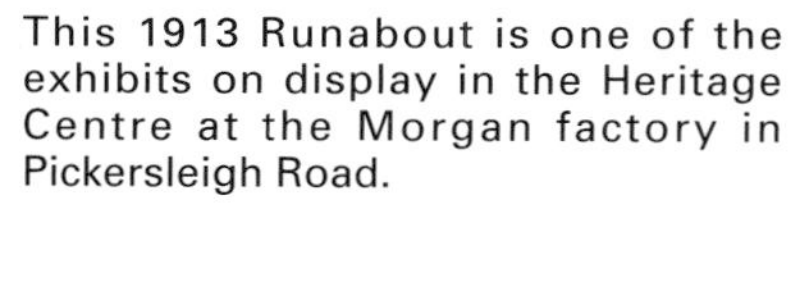
This 1913 Runabout is one of the exhibits on display in the Heritage Centre at the Morgan factory in Pickersleigh Road.

The first 4-4 was nearly 20 years away, but this prototype Morgan, with HFS at the wheel, might well be described as a 3-4. It was the first three-wheeled four-seater, constructed in 1917 on a lengthened chassis.

covering a wide customer span. Much evolutionary development, too, as newfangled features – electric lights, push-button starters and front-wheel brakes, for instance – cascaded downmarket. It says much for the Runabout's original design that it remained competitive for so long, without revolutionary change. Later three-wheelers of different configuration – single front wheel and two driven backs, as on the Bond Bug and Reliant Robin – were never noted for their stability on corners. The wide-tracked Morgans always were. More four-wheelers overturned at Brooklands than proper trikes (motorcycle combinations excepted). At a Light Car Club Brooklands meeting in 1930, the main event, against four-wheeled opposition, was won by a Morgan.

At the basic, mundane end of the scale came the Runabouts, Standard and Commercial – the latter carrying a billboarded container over the back wheel. The sporting fraternity was wooed by the long-wheelbase Grand Prix.

During the 1914-18 conflict, Morgan remained profitable supporting the government's war effort (mainly munitions)

Little had changed outwardly by the Twenties, although the hole in the front cowl had been enlarged to improve the air cooling and accessibility of the V-twin engine. Most of the survivors from this era are preserved in far better condition than when they first left the factory.

This immaculately prepared Family model from the Thirties was one of many three-wheelers at Mog 86.

The front view of the same car revealing the latest divided front cowl and the mesh inserts which in this instance had been painted red to match the wheels and wing ribs of this otherwise black car.

and making cars on a reduced scale for export. Trike development also continued. In his later interview with *Light Car*, HFS said: "In 1915 I built a four-seater model for myself and my family to get about in...it later sold in large numbers, fulfilling a long-felt want..."

Morgan's big event immediately after the war was the construction of a new, much roomier factory in Pickersleigh Road, Malvern Link, financed out of company profits, not borrowed money. Only after it was extended later did Morgan finally vacate its old Worcester Road site: for some time, rolling chassis were made in one plant, then transported to the other for finishing.

With its new factory, Morgan had the capacity to make up to 50 three-wheelers a week to meet huge demand for sound, economical transport immediately after the war. At the 1920 Olympia show, Morgan made several significant improvements to its three-wheelers, among them a quickly detachable back wheel; equal-length drive chains (still only two speeds); a simplified clutch assembly; larger front wheel bearings; a better handbrake; dynamo lighting, and so on.

In the postwar boom, business flourished and production peaked – at around 2,400 cars in 1923. Buoyed by such strong demand, prices were high and HFS prospered. The Rolls-Royce in his garage was an overt symbol of success and wealth. Morgan's golden era did not last for long, however. In the face of fierce competition, led by Austin's more sophisticated four-wheeled Seven, not to mention numerous trike rivals, Morgan astutely trimmed production and prices to remain competitive. Recalling this decline, Peter Morgan told *Car and Driver*: "He (HFS) cut prices from £250 (in 1921) to £95 (by 1925), because as he always said: 'You must build at a profit, but don't try to make a killing.' He invested the money wisely, and after that it didn't honestly matter whether the cars made money or not. It (Morgan) was an investment company, really."

It was a creed to which Morgan has firmly adhered ever since. Not for HFS the mistake made by BSA of over-estimating the three-wheeler market in the face of four-wheeler opposition: Morgan's trike survived the Great Depression of the late Twenties and early Thirties whereas BSA's, launched in 1929, the year of the Wall St crash, folded in 1936.

Morgan's line-up in 1922 included the £155 De Luxe and £160 Family. Water-cooled engines, JAP, Blackburne, MAG or Anzani, cost extra. The following year, cable-operated front brakes, separately applied by a hand lever, were offered as a £6 option. It was a £140 Runabout Standard, with front-wheel brakes, that beat all-comers, including Austin's Seven, in the Junior Car Club's JCC General Efficiency trial in 1923 (and again in '24), with a consumption of 56mpg and a Brooklands lap speed of 55.71mph. Successes like this boosted sales.

Morgan continued to excel in trials, record-breaking and racing, despite being banned, ostensibly on safety grounds, by the JCC from competing against four-wheelers following a crash at Brooklands. Harold Beart, a well-known Morgan dicer and dealer, finally cracked 100mph in 1924 when his

Formula One followers might not believe it, but this is a Grand Prix cockpit! The model was introduced into the range just before the First World War, in recognition of Morgan's success in international competitions, and remained there until 1926.

Another view of Chris Booth's beautifully turned out Grand Prix model showing the full-diameter wheel discs, an impressive display of five lights and the simple but no doubt effective bulb horn.

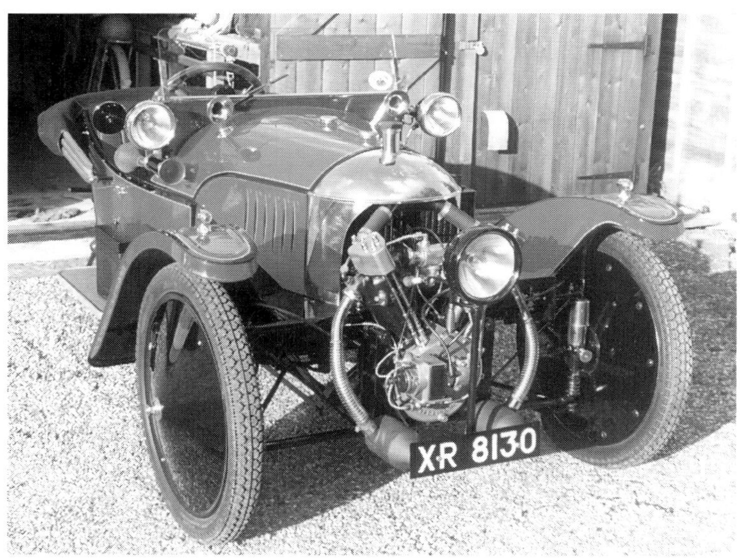

Blackburne-engined Beart Special did 104.63mph over the flying-kilometre, assisted by 'teardrop' streamliner bodywork. At the time it was the world's fastest unsupercharged car. Beart also offered Morgan buyers the luxury of a foot throttle (factory cars still had hand levers, which made gear-shifting tricky), and a column reduction gear to make the steering less direct.

Without successful competition backing and a pragmatic marketing approach, Morgan might not have survived the Twenties, never mind the Thirties, even though its trikes were well designed and made. Best of all, though, they were terrific value for money. When the Austin Seven cost £165 in 1924 (and a Rover 8 £160), you could buy a Morgan Popular for £110. Even the rakish Aero cost only £148 (£160 for a single-seater racing version with overhead-valve engine). Following on from the Grand Prix model, the stark, slim-tailed 70mph Aero – occupants were protected by small aero-screens – came to epitomize the sports Morgan of the Twenties. As press tested in 1926, an ohv Blackburne-engined Aero accelerated from 10 to 30mph in

Bred for competitions, the Super Sports Aero was effectively the successor to the Grand Prix. Note the exhaust pipe with the regulation Brooklands silencer.

second (top) in 5.2 seconds and did 72mph all-out. More impressive still, it would stop in 26 yards from 40mph.

Prices tumbled still further, to under £100 in 1925, though customers had to pay extra for some pretty basic essentials: £10 for a Lucas dynamo lighting set, £6 for front brakes, £2 10s for wheel discs, £4 for a speedometer, £2 for fat Dunlops. Evolutionary improvements included grease-gun lubrication, 7-inch front brakes and strengthened transmission – still two-speed chain drive. Electric starters were introduced as an option in '26. The base Standard had dropped to £89 in 1926, and to £85 in '28. You could then buy a tuned 1,096cc JAP-powered Super Sports for £155.

By 1929, Morgan three-wheelers had been in production for 18 years, steadily evolving, before they were given long-overdue reduction-gear steering that had been introduced as an option the previous year. Racing success still underpinned sales. "Morgan sweeps the board at Brooklands," pronounced a 1929 Morgan advertisement for the Runabout, now costing from £87 10s. Said a *Motor Cycle* report of an 1,100cc ohv JAP Super Sports: "The joy of a high power-weight ratio and the instant surge of power...even when one is travelling at 50mph, provides a fascination which never fails." In its introduction to a 10-40 Aero test in 1930, *Light Car* said: "There always has been something altogether delightful about a Morgan which defies definition...a Morgan is a Morgan, and it can be compared with nothing else but another Morgan." Quite so.

Three-speed transmission came in 1931 – all previous Morgans were two-speeders – followed by a commonized chassis, interchangeable Dunlop Magna wheels and a modern dry-plate clutch instead of the fierce old cone affair that Morgan had used for the previous 21 years. The front half of the drive – engine, clutch and driveshaft – was basically unchanged. However, the bevel-box that turned the drive through 90 degrees on the two-speeder (which continued briefly to special order) was displaced by a conventional three-speed-and-reverse gearbox mated to an integral worm drive. This carried on one side a fabric

A JAP-engined Aero, a model which was introduced into the Morgan range in 1920 and ran through with periodic updates of specification until 1933, this being one of the later examples.

toothed cog that drove the dynamo, and on the other side a sprocket carrying a single chain. The size of the rear-wheel sprocket determined the final-drive ratio. Morgan prices in 1932 ranged from £95 for the hilariously dumpy side-valve Family to £145 for the Super Aero "with specially tuned ohv engine."

Light Car said of a Sports two-seater model in '32: "The middle ratio is very well chosen, for really steep hills can be ascended at 40mph or more." It did not record whether the new 60-degree V-twin JAP made specially for Morgan was any smoother than its 50-degree predecessor. In 1932, the four-model line-up – Family, Sports Family, Sports two-seater and Super Sports – was distinguished by new bodywork and a commonized chassis that lowered the lesser models and slightly raised the SS.

Right from the start, Morgan had used proprietary engines, mostly V-twins – air-cooled and water-cooled, side-valve and ohv – to power its lightweight trikes. One Super Sports special was even fitted with two 596cc water-cooled Scott two-strokes, mated together by a system of sprockets and chains. This 60bhp machine was said by a *Light Car* report to have been exceptionally smooth and flexible. It must also have made a wonderful noise.

In the mid-Thirties, Matchless V-twins – the lovely MX4 yielding over 40bhp – displaced the JAPs that had been Morgan's mainstay supplier from the start. Fittingly, it was with a super-tuned JAP-powered car that Gwenda Stewart rewrote the 1,000cc three-wheeler record book, clocking over 115mph for the flying-kilometre on her way to a number of long-distance records in 1930. Such exploits kept Morgan in the limelight – and profitable.

The final chapter in the Morgan three-wheeler's development story came with the introduction in 1933 of a four-cylinder Ford side-valve engine, nominally of 8hp (933cc), from the Y-type. HFS later described this F-type model as "the best all-round three-wheeler we had turned out." Although Morgan's trike layout had been flattered by much imitation, it was by no means the first three-wheeler

One of the most highly coveted three-wheelers is the Super Sports from the Thirties, a model identified here by its high-mounted exhaust and a spare wheel neatly recessed into the rear bodywork.

manufacturer to employ a four-cylinder engine. Castle Three had done so in 1921, d'Yrsan not much later, and Sandford launched its quality 1,075cc ohv Ruby-powered three-wheeler, capable of 90mph, in 1923.

Morgan's four-cylinder trike was carried by a new, extended Z-section F-type chassis frame – of pressed steel rather than tubular construction – which retained the propshaft-carrying backbone tube. The new 33bhp Family 'four', longer in wheelbase and wider in track, did not displace the V-twins, but supplemented them as a smoother, more refined alternative with four-seater bodywork. Later on, two-seater versions, including a 10hp (1,172cc) model, were added to the range. By now, there was also a foot brake operating on all three wheels. According to *Light Car* in 1936, the F-type 'four' "will slide before it dreams of capsizing." The magazine's tester was impressed with the car's acceleration (a tardy 24.6 seconds for the standing quarter-mile) and top speed (65mph). For *real* performance, the lusty V-twins were still the models to have.

Close-up of the front of the same car displaying its Matchless V-twin water-cooled engine. A similar model was also available from 1933 onwards with a four-cylinder Ford engine.

The Ford-powered models, appropriately called F-type, had extended front bodywork behind a conventional radiator and cowl. This example has returned to Brooklands, the scene of so many Morgan successes in prewar years.

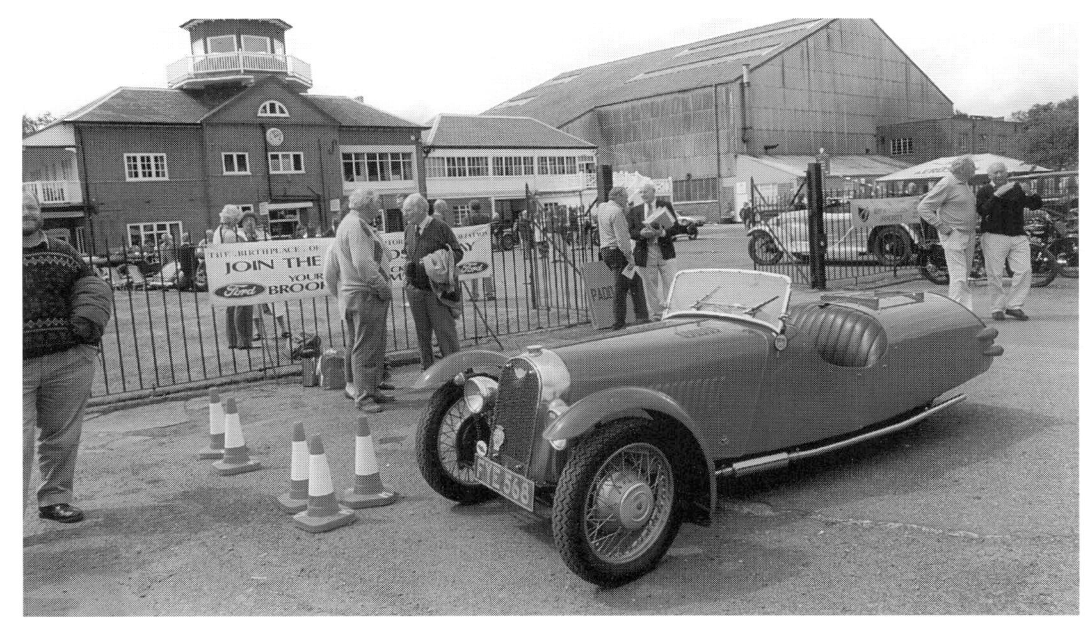

In its test of a Matchless-powered Super Sports, *Light Car* portrayed what Morgan motoring was all about. "To start the engine...requires careful adjustment of the ignition and mixture controls and a few moments work on the crank...the exhaust note, while pleasing to some, can be regarded somewhat gravely by officers of the law and anti-motorists...this is a car that really thrusts the driver in the back...with such widely spaced ratios, a quick change is almost impossible...time lost in closing the hand throttle is amply compensated by the heartiness of the acceleration...once in top gear the car is impatient of restraint." So much for performance. *Light Car* was equally enthusiastic about the handling. "On fast bends...there is no tendency to slide, or for the tail to swing...a flick of the wheel is sufficient to maintain a straight course over the roughest of surfaces." In summing up, *Light Car* used these words: "At £136 10s, the Super Sports Morgan...is a real sports car combining high performance with economy."

Morgan certainly experienced hard times in the Thirties, but through good housekeeping and sound business acumen, it survived less flexible competitors. Sales were not helped when, in 1935, taxation changes reduced the fiscal advantage of running a trike rather than a four-wheeler. As Britain clawed its way out of the Depression, it was obvious that Morgan needed to progress beyond the F-type, which took the trike about as far as it could go. What was needed was another wheel.

The three-wheelers did not cease altogether with the timely launch of the 4/4 in 1936, though their end was in sight, witness dwindling sales – 659 in 1934, 286 in '35, 137 in '36. By 1939, with the storm clouds of war gathering over Europe again, three-wheeler sales were down to 29. Production ceased altogether during the War, but was resumed briefly, on a small scale, immediately after it. In 1949, a *Light Car* leader pleaded the case for low-tax three-wheelers. "We are as certain as we were nearly 40

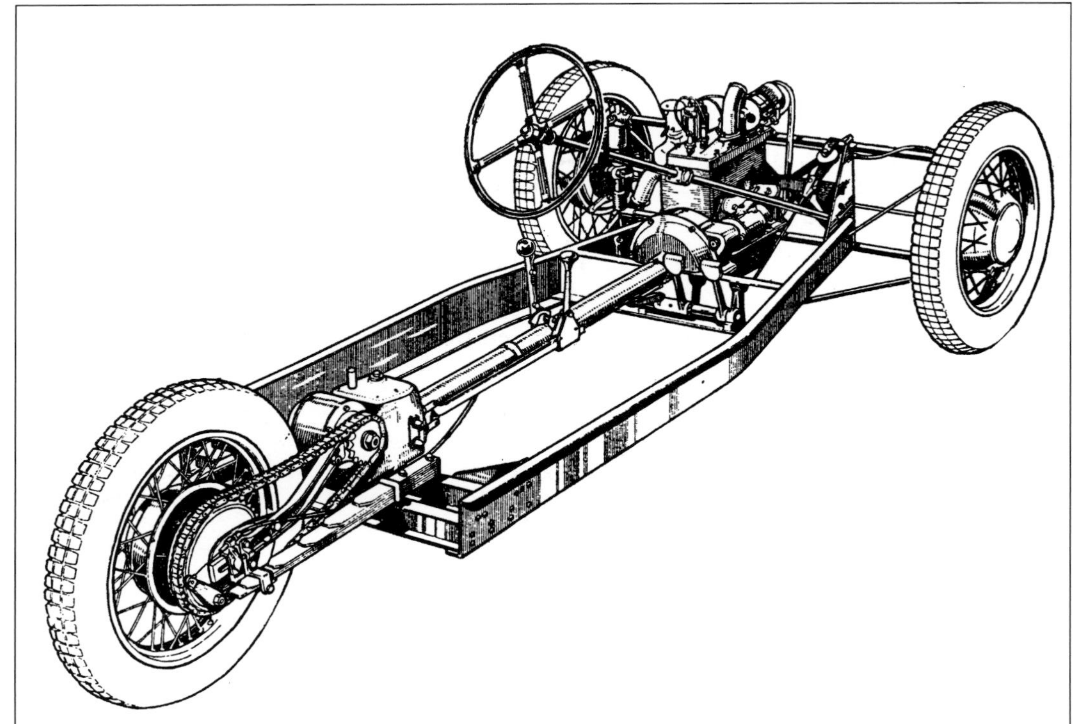

The chassis frame and layout of a Ford-powered Morgan from the mid-Thirties. Unlike the tubular chassis of the twin-cylinder models, the F-type had pressed steel chassis members. The side-valve Ford engine was coupled to a rear three-speed gearbox.

years ago that there is a very definite place for three-wheelers..." But *Light Car* (soon to cease publication) got it wrong. Morgan had outlived its tricar rivals – AC Sociable, BSA, Scott, New Hudson, Raleigh, Reynolds, Economic, Castle Three, LSD, MB, Coventry Victor, Omega, Stanhope, TB and others – through good design and sound marketing. But in the early Fifties it had to export to ensure supplies of steel – and the tax-favourable three-wheelers were a peculiarly British institution unwanted abroad. The last one was made in 1951.

CHAPTER 3

The first four-wheelers

Coventry Climax and Standard-engined 4/4s

Morgan's first 4/4 – four wheels and four cylinders – was described in *Light Car and Cyclecar* late in 1935, after a summer's testing that took in Brooklands without fear of being 'scooped' by a fawning press. The magazine came out just in time to publicize HFS' run in the new model on the London-to-Exeter trial of December 28, '35.

Contrary to popular belief, the 4/4 displayed to a wider audience at the London and Paris shows the following year was not the first four-wheeled Morgan. In July 1931, *Light Car* carried a report of a JAP-powered competition Morgan 'quad' built specially for dirt-track racing. Substituting for the rear fork carrying a single rear wheel was a live GN axle located by radius rods. Apparently it all worked rather well. In December 1932, the same magazine described another hybrid four-wheeler conversion by Sydney Allard (later a car manufacturer in his own right) that mated the front end of a JAP-powered Morgan trike, complete with sliding-pillar suspension, to a unique hotchpotch rump.

Morgan itself had made a four-wheeler back in 1914, when its three-wheelers were banned from certain racing events. Although this car, powered by a four-cylinder Dorman engine, foreshadowed Morgans to come, it was 22 years before anything remotely like it went into production.

You can see in every Morgan made today the legacy of the first 4/4 – written as 4-4 before the War and referred to now as the Series I, though it never officially carried such a designation. It's evident in the car's sliding-pillar front suspension, inherited from the tricars, and the classic long-bonnet, short-tail styling. The simple labour-intensive construction, based on a flimsy cross-braced steel chassis (originally made by Rubery Owen), has changed equally little.

In plan view, the Z-section, cross-braced side rails formed the shape of a boat, pointed at the prow, parallel from behind the engine bay all the way aft. There was no kickup to clear the simple leaf-sprung back axle, originally controlled by Hartford friction dampers; the chassis rails simply ran beneath it. The forward ends of the leaf spring clusters were bracketed to the chassis; the trailing ends passed through slots located in a cross-tube, called the trunnion tube, which also carried the support for the two spare wheels.

Up front, slung across the chassis, there was an upright X-braced framework, stiffened by tie-rods, to carry the independent front suspension – a sliding-pillar arrangement just like that used on the trikes. The framework's upper and lower cross-tubes carried stout spindles or kingpins. In unit with each wheel's stub axle was a sleeve or pillar which slid and swivelled on these kingpins in two phosphor-bronze bushes to provide suspension and steering movement. The pillars acted upon tall, exposed, narrow-section coil springs, which made the suspension extremely stiff in bounce and roll. Vertical travel was strictly limited, metal-to-metal friction minimized by a reservoir of grease, injected through a handy nipple (from 1950, the front uprights were to be press-button lubed with engine oil). Smaller, softer coils (substituted very briefly in the Thirties by rubber bushes or 'snubbers') at the base of the kingpins provided rebound

This prototype four-wheeled Morgan was built up on a modified F-type three-wheeler chassis, although it was soon clear that the Ford Eight engine would be insufficient to power a production car adequately.

The solution was to use the 1,122cc Coventry Climax engine coupled to a four-speed gearbox. This is an early disc-wheeled production two-seater, for which the standard colour was British Racing Green.

cushioning. Morgan also provided a couple of parallel Newton hydraulic dampers to help with suspension control. These dampers were temporarily dropped after the War, following a change of spring rates.

This flexible chassis, originally unprotected and therefore very vulnerable to corrosion, supported a wooden floor, resting on the lower flange of the Z-shaped chassis rails, and a frame of seasoned ash (from Belgium in the early days), made up with the help of formers from dozens of cut and shaped component parts that were glued, jointed and screwed together. As each and every frame was different, the steel bodywork and doors attached to it had to be hand-tailored to fit. Today's Morgans are made in much the same way.

Early 4/4s had 17-inch, four-bolt steel disc wheels known as Easicleans on account of their smooth 14-hole rims and chrome finishers (16-inch discs without holes, standard on the later drophead coupe, were an option). Most cars would leave the factory on Dunlop tyres. Braking was by cable-operated Girling drums, 8-inch diameter in the

Thirties, 9-inch from '46. The handbrake had a 'fly-off' action; after pulling back to release the ratchet, the lever would fly off under spring loading.

The first 4/4 had very crude steering, with a reduction gear mounted halfway down the spear-like steering column; there were no Ralph Nader safety crusaders in the mid-Thirties to warn against such commonplace dangers. This arrangement soon gave way to a chassis-mounted Burman steering box that required under two turns of the 16-inch Bakelite-rimmed steering wheel to go from lock to lock. (When 4/4 production resumed in 1946, the steering system had gained a damper in the form of a saw-steel blade linking each kingpin to the chassis. The effect of this was to reduce, if not totally eliminate, steering wheel wobble.) Among the list of options was a Brooklands four-spoke steering wheel, secured by a brass nut. The column protruded from a cut-out in the wooden dashboard, flanked by gloveboxes – well named in those days as owners really did keep gloves in them. The only cockpit warmth was that received from the engine and exhaust.

Twin spare wheels neatly recessed into the tail panel of an early 4/4 were an indication of Morgan's trials heritage, a form of motorsport in which punctures were common.

This 1939 4/4 two-seater is pictured at St Clair, Dunedin, having been shipped to New Zealand's south island by its third owner. Ray Siatkowski, who describes himself as the car's fourth owner/caretaker, has named the car 'Prudence', after Prudence Fawcett, who achieved fame at Le Mans with a borrowed 4/4 in 1938.

The drophead coupe version of the 4/4 first appeared in 1938 and was reintroduced after the War, powered by a 1,267cc Standard engine. It still retained the rear-hinged 'suicide' doors, whereas those of the roadsters had long since been changed to safer front-hinged ones.

All the switchgear and speedo-dominated instruments (white on black initially, later the other way about) were mounted on a central steel black panel that carried a combined lights/ignition switch, a pull-out starter, a mixture control choke and a big horn push, among other things. These switches could be either white or black, according to what was to hand at the time. There was no rev-counter. No semaphore indicators, either, except on the later coupe, which had high sills that rendered hand signals tricky.

Protection from the elements then (as now) was crude. The roadsters had a simple metal frame attached by screws to body woodwork. You pulled it into position, then cloaked the sticks with a separate black hood, attached to the body with numerous fiddly fasteners – a wretched job in cold, wet weather. The small, flexible rear window gave some aft visibility, but there were no quarter-lights to relieve awful over-your-shoulder blindspots. Fiddly steel-framed sidescreens completed pretty inept weatherproofing. It's unlikely that any of this prewar headgear survives in its original state. And just as well, perhaps. Morgans are best driven fully *al fresco*, even in poor weather.

Before the war, and for many years after it, Morgans were not only hand-made, but hand-painted, traditional coachwork style – fully assembled with a brush. The finish was pretty awful, but the shape was awfully pretty. Who could have predicted, in the late Thirties, that the 4/4's styling would become a design icon of the Nineties, that the basic shape of Morgan's first four-wheeler would make it into the 21st century?

One early bodywork change, *circa* 1938, was to replace the outside spring-loaded bonnet catches with more elegant chrome-plated knobs. Another was to supplant the original rear-hinged 'suicide' doors, fitted to the first 50 or so cars, with safer front-hinged ones that wouldn't fly open under body flexing mid-bend.

The 4/4's original two spare wheels, mounted vertically on the tail, catered for puncture-prone tyres, especially on rough-road trials. The outer arm of the T-bar alloy clamp carried Morgan's winged badge, also found on the flat

This is a publicity photograph of the 1938/39 open four-seater with the Easiclean perforated wheels and central 'dip' or 'passing' light.

radiator grille that distinguished these early 'flat-rad' four-wheelers. Ahead, a horizontal grille bar carried two Lucas headlights and a central 'dip' or 'passing' lamp that came on when the mains were extinguished – by a floor-mounted switch. Modern 12-volt electrics, using two 6-volt batteries connected in series, were a feature of the 4/4 range from the start. On the two-seater roadsters, the batteries were mounted in the luggage bay, adding weight to the back and doing traction no harm. In the four-seaters and coupe, the batteries were under the bonnet.

The first production 4/4, chassis number 3, was delivered in March 1936, priced £185. External door handles cost extra (amazingly, they still do, upholding a Morgan tradition). The tricar's 933cc side-valve engine, with which the first 4/4 prototype (later destroyed in a crash) started life, was considered inadequate for the job. Instead, power came from the well-proven 1,122cc Coventry Climax – Triumph used a similar engine in the Gloria – with pushrod-operated overhead inlet valves and side exhausts. As *The Autocar* observed at the time: "Since some 90,000 Coventry Climax engines have been made to date...reliability may be taken for granted."

As with many other contemporary engines, the RAC's 'horsepower' tax encouraged a much longer stroke (90mm) than bore (only 63mm). With Solex sidedraught carburation, this tall, in-line 'four' yielded 34bhp at 4,500rpm. In the Thirties, when traffic jams were rare, the Climax engine's pumpless thermosyphon water-cooling was normally adequate. These days, the addition of an electric thermostatic fan is considered desirable from a practical standpoint, if not an aesthetic one.

Even though the 4/4 weighed only 13cwt (660kg) ready for the road, its power-to-weight ratio was nothing very special by Morgan three-wheeler standards. Still, a eulogistic press raved about the car's lively performance. Early trials successes in the 'Exeter' and 'Lands End' endorsed the car's competitive edge.

Four wheels, four cylinders – and four gears, too, for the first time on a Morgan. The company's original 1914 four-wheeler had two-speed chain drive, just like the trikes, but not the later production 4/4. Drive passed from the engine-mounted Borg and Beck clutch via a shaft within a rigid tube linking as one assembly the engine and mid-mounted, part-synchro Meadows gearbox. In this position, there was no need for a remote-control shift. The lever sprouted from the top of the 'box and, as contemporary road-testers observed, fell readily to hand. The back-to-front gate, with first and second to the right, and third and top to the left, was not considered a handicap.

The conventional live rear axle, suspended by underslung leaf springs, was driven from the Meadows gearbox by a normal propshaft with needle-roller universals, and a Salisbury spiral-bevel final drive in a 'banjo' casing. Synchromesh was available on third and top, leaving first and second to exercise the driver's double-declutching skills. Late in the Thirties, the Meadows gearbox was replaced by one sourced from Moss, with the same shift pattern, but slightly altered ratios giving different intermediate maxima. Top speed was about 75mph.

A works-entered Le Mans Special – one of seven such cars built, and powered by a 1,098cc Coventry Climax engine – competing over tyre-torturing terrain during a Welsh rally.

A pair of 4/4 two-seaters about to do battle in a sprint with their screens folded to reduce wind resistance.

It takes nothing more than a set of bumpers and tall overriders and a pair of optional exterior door handles to change the appearance of a 4/4.

For all its innate deficiencies, Morgan's patented sliding-pillar front suspension did at least keep the wheels pretty well vertical to the road when cornering and braking, maximizing the grip of the tyres while they remained in contact with *terra firma*. In the 4/4, the assembly carrying it was braced more securely than it had been before to the chassis – described by The *Light Car* as Z-section side members cross-braced by inverted U-section members.

The pretty ash-framed bodywork, long (and generously louvred) in bonnet and no more than thigh high, was fabricated by hand. How else? Little cutaway doors facilitated hand signals, never mind getting in and out. The low seating position and narrow footwells did not. Sounds familiar? It should. Today's Morgans are much the same. Anything you couldn't fit between the wheels, 92 inches apart, stayed at home because most room behind the seats was occupied by a 9-gallon petrol tank and that brace of spare wheels.

In its 1936 road test, *The Autocar* said of the new 4/4: "This Morgan is outstandingly steady, and corners exceptionally well, has light, accurate steering, extremely effective braking, and a lively, willing performance." This test must have been conducted after Morgan had addressed various snags on the original car: engine cooling was improved, the steering column reduction gear displaced by a normal Burman box, and the front suspension modified to introduce some bump 'compliance' by leaning the uprights back slightly. Result: better 'give' and therefore less jarring. Tilting the kingpins (increased in diameter to an inch for production models) gave more favourable castor, too. Mounting the backplates of the Girling brakes more rigidly made for smoother braking.

Early under-bonnet modifications included separating the distributor – its chain drive gave way to a belt – and dynamo, in line, as *The Motor* put it, with modern trends. To cut vibration and harshness, Morgan also thinly insulated the engine from the chassis with rubber bushes.

The 4/4 two-seater roadster was an instant success and soon outselling the rapidly fading trikes. Nearly two years on from its launch, by which time Stewart Sandford,

Morgan's Paris agent (and a manufacturer in his own right), was assembling locally-bodied 4/4s in France, the two-seater was supplemented by the four-place derivative. This might have been called the 4/4/4 – four wheels, four cylinders, four seats – but wasn't. Rather than extend the wheelbase, Morgan ditched one of the spare wheels and squared off the tail, doing little for the car's appearance, but usefully increasing accommodation.

The phrase 'two-plus-two' had not been coined in 1938, but that's what the four-seater Morgan effectively was, space in the tunnel-divided rear compartment being very restricted. The two-seater had a fixed bench-style seat with a common sprung backrest linking inflatable cushions – described as 'floating on air' in Morgan's ads. Seat adjustment was made by pressurizing the cushions (most have since been replaced by plastic foam) to suit the driver's height and weight.

The four-seater, built on the same 92-inch wheelbase chassis, had individual bucket front seats fixed to the rubber-matted wooden floor, so if you didn't fit, too bad. They did, however, have tipping backrests so that

A high-angle shot of the 1938 Coventry Climax-engined 4/4 four-seater owned by Peter Rogers, showing the extensive bonnet louvring and the cover that shields the hood irons around the outside of the rear seat.

An ex-works short-chassis 4/4 built for trials work and subsequently owned by Roger Comber. Like the Le Mans Specials and Replicas, this is one of the most highly coveted early four-wheeled Morgans.

short-straw passengers could clamber in behind. You sat tall in the back, perched above the rear axle, so ride comfort – awful at the best of times – was nigh on intolerable. Like the roadster, the four-seater had a fold-flat, metal-framed screen, a simple removable hood and separate sidescreens. All this on an unchanged chassis for £225 – £15 more than the basic two-seater.

In the summer of 1938, Morgan actually reduced its prices, to £199 10s (two-seater) and £215 5s (four). In at the top came another (two-seater) variant – the elegant £236 5s drophead coupe. Based on an Avon prototype, this car was distinguished by its elegant sloping tail and spare wheel cover. With its fixed, wood-framed screen, full-height doors – always rear-hinged 'suicide' affairs – sliding glass windows and three-position hood (in place of a simple canvas canopy), the drophead looked altogether smarter, more upmarket than the rakish roadsters. Its electric wiper motor (nothing so crude for Morgan as a vacuum pump) was out of sight, below the wooden scuttle, working clap-hands blades. In the roadsters, the motor sat in one corner of the screen, working the opposite wiper by a tie-rod.

Two seats or four, the 4/4's sporting character was underlined by its low build, 90mph speedometer (*The Autocar*'s best quarter was actually 78.3mph), and short-throw gearlever. By its success in motorsport, too. With the so-called Competition engine – a small-bore 1,098cc version that qualified for racing's popular 1,100cc category – boosted by go-faster bolt-ons, the little Coventry Climax engine was good for over 50bhp, perhaps 60bhp in cycle-winged racing trim. It was such a car, specially prepared and lightened, that finished second in class and 13th overall (out of 42 starters) at Le Mans in 1938.

Without Morgan's assistance, the car was entered and co-driven by Prudence Fawcett, whose cavalier have-a-go drive in the 24 Hours was said to have been her only serious foray into motor racing. The success of Fawcett (who died in 1990) inspired Morgan to build an 80mph, £250, cycle-winged Le Mans replica. Only three were made, though according to Morgan historian Ken Hill in *Completely Morgan*, there were also seven Le Mans Specials and four TT Replicas as variations on the 1,098cc theme. All are today among the more coveted of Morgans.

Just as noteworthy as this Le Mans success was George Goodall's class win in the RAC rally for three consecutive years, in 1937, 1938 and 1939. Goodall had joined Morgan

One of just three Le Mans Replicas built prewar but not sold until the late-Forties. This immaculate example was owned for many years by Morgan historian Ken Hill.

A Morgan chassis and an estate car body may seem an unlikely combination, but this example by Peamore Garages, of Exeter, is a bold attempt to produce an acceptable result. A load space of 40cu ft was provided behind the doors.

in 1925 and was promoted from general manager to managing director in '37 when HFS, distancing himself from the company's day-to-day affairs, assumed his late father's role as company chairman. The Reverend George Morgan had died the previous year, aged 86.

Before the War, there was one further development of great significance for Morgan: a new engine. Supplies of the 4/4's original Coventry Climax unit were threatened by the financial woes of Triumph, who supplied Morgan with its made-under-licence Climax engines. Although Morgan experimented with an Arnott-supercharged Ford side-valve engine, it chose another ohv design developed specially for the 4/4 by the Standard Motor Company (not yet in bed with Triumph). This was at the express wish of its head, Sir John Black, rekindling an old association: nearly three decades before, remember, it was young Black who had produced for HFS the drawings to back his successful patent application for sliding-pillar front suspension. Standard was already supplying engines to William Lyons' blossoming Jaguar (*nee* SS) company. Why not Morgan, too?

The Standard Special engine, effectively an overhead-valve conversion of the Standard Ten's side-valve 'four', first powered a Morgan (George Goodhall's hack drophead coupe, to be exact) in 1938. Production models followed in '39. To minimize RAC horsepower tax, its 100mm stroke was even longer than that of the displaced Coventry Climax unit, the bore only slightly bigger at 63.5mm. This gave a displacement of 1,267cc and an output of nearly 40bhp (up 15 per cent on the displaced CC unit) at 4,500rpm. The engine would rev to five-five, which was pretty racy in the Thirties, and thermosyphon cooling was assisted by a two-bladed fan. Postwar cars got a water pump, too.

Initially, this tall new engine, which had Morgan's name on the rocker cover, was standard on the drophead coupe, £5 extra on the open sports. Few of either, though, were made before the War intervened and the Pickersleigh Road factory was turned over to more pressing matters, not least the development, in the leased-out woodshop and mill, of an in-flight refuelling system for the RAF.

Discounting incomplete cars that left the factory as rolling chassis, Morgan made just over 1,300 Series 1 4/4s (chassis numbers, allocated according to orders, stretch to over 2,000, but don't be fooled by that) in the war-interrupted 1936-50 period. All but a handful had Coventry Climax engines, and most were painted green.Fewer than 40

CHAPTER 4

The postwar 4/4s

Establishing the Ford connection

Standard-engined 4/4s (originally 4-4s, remember) were made in the Thirties before car production at Malvern Link was interrupted by war. As steel supply was linked to export sales after the conflict, Morgan drummed up business abroad by appointing agents in the US, Canada, South America, Australia and South Africa, as well as in Europe. (Today, as a legacy of this postwar expansion, Morgan still sells over half its cars abroad, though it is no longer heavily dependent on any one market).

In its July 5, 1946 issue, *The Autocar* described the 4/4 as if it were a new car, though little had changed since 1939. The pushrod ohv engine, nominally rated for taxation purposes at 10hp, was slightly bigger – and 15 per cent more powerful – than the Coventry Climax unit it displaced. Being fashionably long in stroke and narrow in bore, it was also very flexible. With a single Solex downdraught carburettor and modest 6.8:1 compression ratio (postwar pump petrol was a low-octane brew), the 1,267cc engine yielded 40bhp at 4,300rpm.

It was a tough little engine that Standard made for Morgan, and quite an efficient one by the yardsticks of the day. Its vertical valves, operated by long pushrods, worked in high-turbulence combustion chambers in which the spark plugs were set horizontally. The engine was mounted in the Z-framed, floorboarded chassis more flexibly than in previous Morgans, though the transmission was like the earlier 4/4's.

Up front, in the engine's bellhousing, was a single-dry-plate Borg and Beck clutch. As in the original 4/4, the four-speed, mid-mounted gearbox was separated from the engine – but rigidly connected to it – by a stout tube enclosing a short driveshaft. This arrangement gave several benefits: it created more room for big feet; the gearlever operated directly on the 'box, not through a corrupting remote-control linkage; and it shortened the propshaft (and therefore lessened the risk of whip vibration) to the underslung rear axle. It also gave a better weight distribution. Agile handling, remember, was central to Morgan's creed. Using machine tools acquired during the War for other purposes – parts for the Oerlikon anti-aircraft gun, as well as aircraft undercarriages, were made at Pickersleigh Road – Morgan was in good shape to make in-house components. Any thoughts of making its own engine, though, had been abandoned before the War. "Father thought about it," Peter Morgan revealed in his interview for this book. "He got a friend, an engineer with AJS, to do a four-cylinder ohv design...we still have the drawings." Nothing came of them.

With the new engine came modifications to the sliding-pillar front suspension. By changing the spring rates, particularly those of the bottom rebound spring, Morgan temporarily did away with the Newton hydraulic dampers (not shock absorbers – it was the springs that absorbed the shocks). Also new was the saw-steel damper blade, linking each kingpin to the chassis, used to counter a tendency for the wheels to wobble. The 8-inch Girling drum brakes were applied by cable – hydraulic actuation was still some way

This immaculately presented 1956 Series II 4/4 two-seater owned by Robin Thompson is the result of a ground-up rebuild completed in 1996. The car has a Ford 100E engine.

The only non-standard items visible here on YKK 678 are the rear overriders from a later car.

off. With only around 14cwt to stop (15cwt for the drophead coupe) the brakes were not exactly overtaxed.

Before the War, in 1939 you could buy a Standard-engined two-seater 4/4 for £199 10s, a four-seater for £215 5s and a drophead coupe – a truly elegant car – for £236 5s. When Morgan resumed business in 1946, prices had more than doubled, to £454 15s 1d (with purchase tax) for the two-seater and £505 18s 3d for the drophead coupe. Inflation was well into its rampant spiral.

The Autocar was impressed with the drophead coupe it tried in '46. "The driver finds himself sitting well down inside what appears to be a high-sided body, but the forward visibility is good...the steering is fairly quick and the car feels extremely safe at speed and can be placed accurately. The suspension is a nice blend of stability with comfort...there is no suggestion of sponginess. There is plenty of life in the engine and the gearchange is a quick, snappy one and the brakes are powerful. It is not a luxury vehicle, but a lively and satisfying small car for the fan who drives for the sake of driving, enjoying every minute of it."

Mechanically, Robin Thompson's 4/4 is all correct to original specification, including the Ford side-valve engine and three-speed gearbox.

Close-up of another Series II 4/4, this one of 1958 vintage, showing the later rear light arrangement and, barely visible, a Wooler remote gear linkage.

Postwar petrol rationing prevented the Standard-engined 4/4 from upholding Morgan's competition record, though the Morgan team triumphed in the foreshortened 'Lands End' in '49.

Although it had been around for over a decade, largely dormant throughout the War, the Standard-engined 4/4 was in production terms only four years old when it was pensioned off. The decision to axe the car was forced upon Morgan by Standard's new one-engine policy which meant the chop for the little 1,267cc pushrod 'four'. The blow was softened, though, by the availability of a superior replacement – the Vanguard's engine – just when the market was ripe for something a bit beefier.

The story of the Plus 4s unfolds in the next chapter. Here, to continue the 4/4's history, we must jump five years and a spurned takeover of Morgan by Standard. "I was there, at Earls Court," recalls Peter Morgan, "when Sir John Black asked for a word. He wanted father to come in with him. He promised me a job." Fortunately, the approach was rejected: neither Morgan, father nor son, liked the idea of

This works demonstrator 4/4 was frequently driven by Peter Morgan in trials, but for a change it was given a smoother ride during a day out at the Guild of Motoring Writers' test day at Goodwood.

John Homfray's 1962 Series IV 4/4 looks virtually original apart from the non-standard spotlamp and seat belts and the temporary steering wheel (fitted to placate an MoT tester), but its original Ford 1,340cc engine had been replaced by the later 1,498cc GT version.

There is a great enthusiasm for Morgans in Germany, where this 1969 4/4 four-seater has ended up in the ownership of Herr Wiegand. Note the neater tail treatment compared with earlier four-seaters.

An under-bonnet view of Herr Wiegand's '69 4/4, showing the fabricated exhaust system and electric cooling fan.

becoming a small cog in a big wheel.

It was in 1955, when Stirling Moss won the Mille Miglia and Donald Campbell broke the water speed record, that a 4/4 was once again listed by Morgan after a break of half a decade. Although the new steel-bodied base model, introduced as a prudent economy measure, inherited the Plus 4's chassis and radical facelift (the story of the cowled grille and fared-in headlights is told in Chapter 5), the 'modern' Series II (rather than Series 2) 4/4 was in some respects retrogressive, a lesser car than the earlier 'flat-rad' Series I. However, as the cheapest open two-seater on the UK market – it was £638 12s 6d when announced at the Earls Court show in 1955, £713 17s as tested by *The Motor* in 1956 – it fulfilled a need for the impecunious enthusiast, if not the speed freak. Not until the Austin-Healey Sprite arrived was there a cheaper alternative.

Conceding the car's modest performance, Morgan wisely referred to the new 4/4 as a tourer, not a sportscar. There was no ohv engine this time, but Ford's 1,172cc 100E side-valve unit – a development of the 8hp and 10hp engines formerly employed by Morgan in its three-wheelers. The best that could be said of this rather crude and dated engine, albeit stronger and peppier than when Morgan last

There is no disguising the fact that a four-seater Morgan makes a more attractive picture with the hood down, but this wire-wheeled 1970 model, seen at Hopmog 94, is a well turned out car.

used it, was that it was cheap, reliable and easy to service. It also fitted easily under the 4/4's lowered bonnet – bereft of louvres, to keep down costs, unless you paid extra to have them stamped in. With an output of 36bhp at 4,400rpm, though, it was less powerful than Standard's axed ohv unit.

The good news about the Series II's gearbox was that it was cheaper than the Series I's displaced Moss 'box. It also came mated to the engine. The bad news was that it had only three gears, to the detriment of performance, and would require some sort of remote-control shift mechanism. This took the form of a crude, Citroenesque horizontal push-pull arrangement, pivoted to an extended gearbox lever. The surprise was that it worked so well – much better than appearances suggested. It was no ball of fire, this Series II 4/4 tourer, but it was certainly cheap and its footwells quite spacious.

Even the road-testers of *The Motor* and *The Autocar*, still a bit cautious of overt criticism, were unimpressed by the performance of the Ford-engined 4/4: call it 75mph all-out at best, and 0-60mph in a yawning half-minute. Both magazines were more complimentary about the car's ride and roadholding. So was *Road & Track* in America. Reporting on the 4/4 Competition, fitted with high-compression 8:1 Aquaplane cylinder head and twin SUs (in place of a single Solex), it said: "Extraordinary handling makes up for a lack of cubic inches in this uncompromising classic." With a claimed 44bhp gross, the Competition's top speed increased to 80mph, and the 0-60mph time fell by a full 10 seconds. Even so, it was a pretty sluggish car without more aggressive tuning – available through the go-faster specialists of the day. Many of these early side-valve Series IIs, and later ohv Series III, Series IV and Series V cars, for that matter, were the recipients of engine/gearbox transplants as more desirable Ford powertrains became available. Bizarre though it sounds, a Series II with a second gearbox, doubling up on gears – and gearlevers – was a reality, it seems.

If not one of the more memorable Morgans, the Series II 4/4 was significant for being cheap and rekindling an association between Malvern and Dagenham that was still going strong 40 years later. Tapping into Ford's technology – and a never-ending ladder to bigger and better things – was an astute move. After the 100E came a succession of more notable Ford engines, starting with the new high-revving, short-stroke 105E (Series III) from the new notchback Anglia. Although smaller in capacity at 997cc, it was bigger in heart and output – 39bhp against the old engine's 36bhp. Moreover, there was huge potential for tuning, as the 105E-powered Formula Junior single-seaters were to demonstrate. Anglia racers, too, like Broadspeed's blistering saloons. With the new engine came Ford's four-speed gearbox, necessitating minor chassis mods to accommodate it; narrower wings compensated for a slight increase in body width. The horizontal push-pull gearlever, acting on a stub poking up from the lid of the far-forward 'box, was retained in modified form. Again, it worked surprisingly well. *Car and Driver* quoted in-gear maxima of 22, 38, 64 and 80mph for the Series III – and a 0-60mph

Carrying a coveted number-plate, this 1972 4/4 two-seater belonging to L Alexander is powered by a 95bhp Ford 1600 GT crossflow engine.

time of 26 seconds was recorded.

HFS Morgan didn't quite make his company's 50th anniversary – or the resurrected 4/4's fifth. Since before the War, Morgan's founder had been living in Maidenhead, well away from the Malvern factory, which he visited regularly, usually midweek, driving up Tuesdays, returning Thursdays. In his 1985 interview for *Car and Driver*, Peter Morgan recalls a change in the schedule. "This time he drove up on Friday for my children's half-term. He collapsed on Saturday during breakfast...and he died on Monday. Heart attack...it was the right way for him to go. I'd been awfully worried about how he would be when he couldn't drive...as it was, driving his Morgan all the way up from Maidenhead on the Friday, and dying on Monday, was marvellous."

HFS' death was the end of an era, and brought with it problems for the company he founded. Peter Morgan told Charles Fox: "It wasn't easy, because when he died, the money was divided between us children and the cars suddenly had to make a profit if they were going to continue. I got all the chaps together at the works and told them: 'Right, now we've got to do something to keep going. We've got to make a go of it.' And they did. They were very good about it." In his interview for this book, Peter Morgan said: "I've been very, very lucky in having such an excellent workforce. Although we've had our differences, they've always pulled out the stops when it mattered."

Before the magazines got round to testing the Series III, of which only 59 were made, it had given way, in 1961, to the Series IV, powered by the Ford Classic's 1,340cc 109E engine, giving 40 per cent more power – 54bhp instead of 39. According to *The Motor*, "the new engine... revolutionized performance." As a staffer on the magazine at the time, I can vouch for that. There was a big increase in torque, too, from 53lb ft to 74. With the now familiar push-pull change – you rocked the knob sideways to cut across the gate – came a lot more verve. Something approaching exhilaration, indeed, for the first time in a 4/4. What was described as "outstanding top gear performance" was not all down to increased power, however. Lower overall gearing played its part, too. To cater for a 90mph top speed, Morgan also fitted Girling front discs from the Plus 4 to give exceptional braking for a 13cwt car.

It was a logical step from the Classic-engined 1,340cc Series IV, which was made for 18 months alongside the more popular Plus 4, to its 116E Cortina-powered 1,498cc Series V replacement, which ran from 1963 to 1968 in standard (60bhp) and Competition (78bhp GT) forms. Along with the uprated engine came an improved, all-synchro four-speed gearbox – still with the crude push-pull shift unless you opted for Wooler's remote-control linkage. This was rendered unnecessary on later Series V cars by a longer Lotus Cortina gearbox, which allowed the use once again of a conventional shift.

As Ford's mainstream engines grew in capacity and power, so too did Morgan's. With the Mk 2 Cortina came a crossflow cylinder head – induction and exhaust were on opposite sides of the head to improve gasflow – and another hike in capacity, from 1,500 to 1,600cc. Enter, then, the Series VI 4/4? Not so. Breaking with tradition, and confusing Morgan students, the 1,599cc cars launched early in 1968 were called 4/4 1600s.

All production 4/4s built since the series was resurrected in 1955 had been two-seaters, but with the additional power of the 1600, Morgan reintroduced optional four-seater bodywork. There were essentially two versions of the first 1600 engine, one with 74bhp (standard Cortina), the other, fitted with twin-choke Webers, a lightened flywheel and plastic fan, with around 88bhp (Cortina GT). As the quicker version was the more popular, most Kent-engined 1600s came with GT power. Indeed, the lesser engine was eventually phased out through lack of demand, and the 'Competition' tag dropped.

The 88bhp GT-engined 4/4 was tested by *Autosport*'s John Bolster in March 1974. He didn't like the standard seats (individual buckets cost extra) or the ride ("very hard indeed") but he did like the raspy exhaust ("really, Peter") and the performance ("lively, with excellent acceleration"). He quoted a 0-60mph time of 9.8 seconds, and a top speed of 102mph, suggesting that this was the first production 4/4 to exceed the ton.

Many of the improvements introduced for the Plus 8 (see Chapter 6), launched in 1968, were soon carried over to the 4/4 as well. These included revised rear lights, dual-circuit brakes and a better crash-protected facia. More of this later.

With the approaching demise of Ford's 1600 Kent engine, Morgan was in need of a suitable replacement that satisfied various criteria, not least that affecting tightening emissions regulations, which had effectively forced Morgan out of the US, once its best export market. Referring to this decline, Peter Morgan told *Car and Driver*: "One year we sold 87 per cent of our production to America. Then America seemed to go into a slump, and our agent didn't order a car for 11 months...it was tricky."

Ford had a clean-sheet 1600, of course, in the shape of its overhead-cam CVH (compound valve hemispherical) for the new Escort. The problem was that this engine was designed for a front-drive, transverse layout, not a rear-drive longitudinal one. Faced with what seemed like a technical impasse, Morgan turned instead to Fiat's respected 1600, as used in the Mirafiori saloon. Thus was the first twin-cam Morgan launched – discounting those factory specials with experimental Lotus Cortina and Ford BDA engines.

As told by *Motor*, the story – now part of Morgan folklore – goes that a Ford luminary strolled onto the Morgan stand at the 1980 Earls Court show and told Peter Morgan he'd have to cancel his order for a 4/4 if it was delivered with a Fiat engine. Wouldn't he reconsider Ford's CVH, as used in the nippy Escort XR3? He would – and did. Ford, keen to retain its long association with Britain's oldest sportscar manufacturer, devised a hybrid north-south powertrain, using a modified Cortina sump and a special flywheel in a Capri bellhousing mated to a four-speed Ford gearbox. Mounts, bulkhead, radiator and exhaust needed special attention, too.

Morgan customers could thus choose in 1982 between a five-speed Fiat twin-ohc (1,584cc, 98bhp) and a cheaper four-speed single-ohc Ford (1,599cc, 96bhp). Dynamically, the five-speed Fiat probably had the edge – at least until the Ford version, which was the more practical from a spares and service standpoint, gained the Capri/Sierra's five-speed gearbox. Most buyers opted for the Ford.

As it was clearly pointless for Morgan to make two versions of the 4/4 that were so alike in performance, one had to go. Exit the Fiat Twin-Cam, after only 92 had been made, in favour of the cheaper Ford – and the continuation of a long association. The happy outcome of this decision was that, for a while, the Italian affair continued with a 2-litre, fuel-injected Fiat engine that relaunched in 1985, after an absence of 17 years, the Plus 4 (see Chapter 5), nicely bridging the gap between the Ford-engined 4/4 and the Rover V8-engined Plus 8. Morgan has retained this three-model line-up (but not the Fiat engine) ever since.

In its road test of a Ford CVH 1600 4/4 in '82, *Motor* found the top speed of 103mph to be well down on the donor XR3's 111mph; given the Morgan's dreadful

All ready for some action, Roger Bluff's 1983 4/4 four-seater is about to perform in the Editorial Bouncers production car trial at Lintridge Farm. Photographer Cliff Baker, who provided many of the pictures for this book, is in the back

Rostyle wheels became a standard fitment on the 4/4 in the Eighties. This four-seater, which dates from 1983, wears the aluminium bumpers which, devoid of overriders, had been introduced in 1977.

At its best, Morgan weather protection was never up to much. Draughts are guaranteed, even with these floppy sidescreens.

aerodynamics, that's not too surprising. However, Ford's fast hatch was also quicker to 60mph, belying its inferior power-to-weight ratio. Despite "primeval" front suspension, "virtually extinct" worm-and-nut steering – changed as a result of supply problems to a French Gemmer setup in 1984 – and a "crude" leaf-sprung rear axle, *Motor* found the handling good on smooth roads. On bad ones "things are less predictable, the 4/4 getting jarred and even thrown bodily off line by mid-corner bumps. Over sharp undulations, the ride is so bad, it's often necessary to take a firm grip of the wheel..." Quite so. Little had changed. A Morgan's a Morgan.

Engines came and went, but the 4/4, which became more popular when the Plus 4 was temporarily dropped in 1969, was not neglected in other departments. Far from it, especially on the safety and security fronts. In '69 came a new dashboard layout, incorporating a small rev-counter, followed in '70 and '71 by a collapsible steering column, an ignition/steering lock, a multi-purpose column stalk, cockpit crash padding, trim-matching vinyl facia, anti-burst door locks (sourced from Land Rover), stronger door hinges, dual-circuit Girling brakes and 'torpedo tube' rear lights – as on the Plus 8, of course. The next three years saw the introduction as standard of a crude fresh-air heater (which displaced the old fug stirrer option), dashtop demister vents and individual bucket seats (previously extra): legal requirements banished Morgan's traditional bench – and good riddance, too.

Several changes were made for 1977: in came yet another new dash (with the rev-counter now paired with the speedo ahead of the driver, and the supplementary dials in the middle, above a row of rocker switches), out went the old chromed bumpers, to be replaced by overrider-less aluminium ones – now standard at the back. Aluminium became an official bodywork option at this time, mainly as an anti-rust ploy on the 4/4.

For Morgans in general, not just the 4/4, 1986 was a milestone year. It was around this time that quality took a

The hood design and rear-quarters visibility had improved dramatically by the Seventies, but closed to the wind this 4/4 looks ungainly.

major turn for the better with the introduction of epoxy powder coating, galvanization, wings-off ICI painting, Cuprinol dipping, not to mention inertia-reel belts, improved weather protection and, for the 4/4, Gemmer steering. Anti-corrosion measures progressed with passivated nuts, bolts and fasteners. In 1988 came undersealing (previously an option), another upgraded heater (this one could be switched off in the cockpit), and in 1990 the adoption of wire wheels as standard.

After the Plus 8, the Kent-engined 4/4 1600 is the most numerous of all four-wheeler Morgans. Of the 3,512 built from 1968-82, no fewer than 43 per cent were four-seaters; the extra space – usually used for luggage rather than offspring – outweighs the loss of looks for many Morganists, it seems. By contrast, the Fiat 1600 ranks as one of Morgan's rarer models and is greatly outnumbered by the Ford CVH 1600, ordinary and lean burn, displaced by the 1600 EFI (electronic fuel injection) in 1991. It was with this fully mapped, Weber/Marelli-injected 100bhp engine that Morgan entered the world of serious emissions control, unleaded petrol and catalytic converters.

The latest chapter in the 4/4's development, if not the last, centres on Ford's 1.8-litre Zeta twin-cam 16-valve engine, like that used in the Fiesta RS 1800 and Cabriolet. It was introduced in 1993 (no 4/4s were made in '92 because of engine supply problems) and, with 128bhp, promised the best 4/4 performance to date. But as development and production engineer Bill Beck recalls, 128bhp was unrealistically high. "We couldn't reach the claimed power output either on our dyno or that of Bridgend (Ford's engine plant)." In Germany – a strong Morgan market – this was an embarrassment, as buyers there can return cars that don't fulfil their makers' promise. Taking its cue from German manufacturers, wary of exaggeration (hence their pessimistic performance figures), Morgan reduced its claims. Beck says that in January 1997 the 1.8's power really did drop as a consequence of tighter emissions control. When this book went to press, Morgan's brochure listed the

4/4's power as 114bhp at 5,750rpm, torque as 118lb ft (respective figures in a previous brochure were 121bhp and 119lb ft).

To accommodate the Zeta, brought in as a complete engine/gearbox unit from Ford's Power Torque offshoot, Morgan had to alter the chassis. As it happened, lowering the powertrain (which has an awkwardly tall gearlever) had the benefit of straightening out the drivetrain from gearbox (ex-Transit, Sierra) to back axle, reducing wear in the one-piece propshaft's universal joints. "Wear used to be a problem in the front u/j," Bill Beck conceded. Unlike Caterham, when using the same engine, Morgan does not have to relocate the alternator.

When Girling opted out of the small-scale specialist market, Morgan changed to disc/drum Lockheed brakes, introducing for the first time four-pot front calipers and servo assistance across the board. In the interests of rationalization, Morgan standardized on this setup in 1993 for all three models, so the 4/4, with the same anchors as the Plus 8, was if anything over-braked. Unlike the Girlings they replaced, Lockheed's rear drums were self-adjusting.

Again it was problems with supply – to which Morgan has always been very vulnerable – that prompted a change from a GKN Salisbury back axle made in England to an Australian-sourced BTR one that came from the other side of the globe. "We had cars all over the place at the factory without back axles," Bill Beck recalled. Morgan is full of praise for the BTR replacements, which not only cost less but, a year on from their introduction, were free from warranty claims. GKN's rear axles could be very noisy.

Airbags have been on Morgan's agenda since the early Nineties. Much time and money has been spent on perfecting their use in a car that does conspicuously well in crash-testing, even though it was never designed for it. "You never get any footwell intrusion, and the doors still open and shut," says Beck. That means no trapped feet.

When this postscript was written in April 1997, Morgan was preparing, for US distributor Bill Fink, the first production Morgans to be equipped with airbags. Their installation demanded some fairly basic, far-reaching changes which, in the interests of rationalization, were to be applied to all cars, whether fitted with airbags or not, from the summer of '97. These revisions include longer doors (extended back by about 70mm), a relocated heelboard, a dash set 2 inches further in (Beck describes it as "coupe style") and a dished bag-carrying wheel. "We needed to move the driver back," says Beck, acknowledging that airbags can be counterproductive (*ie* dangerous) if they are not an optimum distance from the crash victim's chest. A useful by-product of all this is that there's more legroom for tall drivers; the seat runners are the same length as before, but they've been moved back.

Morgan's new paint shop was fully operational early in '97, after delays with the filtration system. Because of the dust-free atmosphere in which Morgans are now sprayed, the paint finish is better than ever. Pity that Morgans are also quieter than they've ever been before: to meet EU regulations, the 4/4 (and the Plus 4 and Plus 8 for that matter) were slated for new noise-abating stainless steel exhaust systems in '97.

CHAPTER 5

The Plus 4s – *real* sportscars

Triumph power sets a new standard

Morgan has always been at the mercy of its engine suppliers. Peter Morgan, who joined his father in the family firm in 1947, recalls that one of his first major tasks was to find a suitable replacement for the 4/4's little ohv Standard engine that was soon to be axed. "For economy reasons, father didn't want anything above 1,500cc, at the most 1,750cc. Leonard Lord couldn't supply, GM were not interested and we didn't want a Ford as we'd then be totally dependent on Dagenham for engines." The tricar, remember, was powered by a 100E motor.

Peter Morgan didn't have to look very far. Existing suppliers, Standard, appeared to have the answer in the new Vanguard saloon's ohv 'four'. HFS thought the jump from 1,267cc to 2 litres would be too big, but his son won him over. "We fitted the engine into a chassis and I took father for a ride up a local steep hill. 'This is wonderful,' he said of the torque. Father was so impressed, he agreed to the Vanguard engine, even though it was bigger than he wanted." As it happened, bigger than both Morgans bargained for. "To our absolute horror," recalls Peter, it came out at 2,088cc, which meant we would have to race against Aston Martins, Healeys and goodness knows what. It was a big blow."

The ubiquitous Vanguard engine, also used in the Ferguson tractor (and in modified form in the subsequent TRs of Triumph, owned by Standard since 1945), was rated in Britain for taxation purposes at 18hp. With a single Solex carburettor, pushrod-operated overhead valves and modest 6.7:1 compression ratio, it developed 68bhp at a lazy 4,200rpm. Under 70 horsepower may sound puny by modern standards, but it represented a 70 per cent hike in power for the little Morgan. Such a quantum leap in performance justified a new name – Plus 4 – and the resurrection of an old slogan: "fastest at the price."

The layout of the transmission was just the same as in the 4/4, with an Elektron tubular casting rigidly connecting engine and mid-mounted four-speed Moss gearbox (the Vanguard's three-speeder was rejected). To stiffen the chassis, Morgan replaced the channel-section crossmembers with three box-section ones, in addition to a rear bolt-on trunnion tube which slot-located the rear springs. Bulkhead rigidity also came in for attention, with welded sheet-steel reinforcement. To accommodate the bulkier engine, the wheelbase, hitherto 92 inches (2,337mm), was stretched to 96 inches (2,438mm), improving legroom. The track and cockpit were also broadened slightly, and the footwell widened at pedal level.

Lengthening the front suspension's kingpins allowed the use of longer, softer springs to improve ride comfort (which was still pretty awful). Morgan also introduced a semi-automatic oiling system for the front suspension, which needed ample lubrication if it wasn't to stiffen up and wear prematurely. Pressing a button on the bulkhead every hundred miles or so directed a slug of engine oil to the uprights. It was better than nothing, but not as good as grease.

Movement, such as it was, of the front suspension was now controlled by Girling telescopic dampers. At the back,

The original Morgan Plus 4, which replaced the 4/4 in 1950, was built on a 4in longer wheelbase and had a 2in wider body covering the Standard Vanguard 2,088cc engine.

Like the 4/4, the Plus 4 was offered with a choice of two and four-seater roadster bodywork as well as this drophead coupe, which was still equipped with rear-hinged 'suicide' doors.

piston dampers (replacing friction ones) acted on the usual half-elliptic trunnion-located 'cart' springs (this arrangement gave way to conventional shackles in 1955). Four-bolt steel wheels carried 5.25x16 Dunlop tyres – a size bigger than the 4/4's – and 9-inch drum brakes, now hydraulically actuated. Newfangled discs were not used until '59.

The first Plus 4 was available in the usual three forms: two-seater, four-seater and drophead coupe, at prices ranging in Britain from £652 8s 4d to £722 13s 10d, purchase tax included. Although the fastest four-wheeler Morgan yet – *The Autocar* wrote of "vivid acceleration and high natural speed" – it was all-out at 85mph and struggled to better 18 seconds for the 0-60mph dash. Still, by 1950 standards, the Plus 4 was quick. As *The Motor* put it in its 1,900-mile 'continental' road test (the team went to Belgium): "It accelerates in an invigorating manner which whisks it effortlessly past ordinary traffic..." Springy steering that became very heavy on sharp corners – one of the drawbacks of the sliding-pillar arrangement – didn't impress

An interesting one-off design, this four-seater saloon by Cooper Motor Bodies, of Putney, was rebuilt by Gerry Wilburn in America, where it became a prizewinner at the Pebble Beach *concours* before returning to Europe and its latest home in Holland.

The Motor's test drivers, however.

The early Plus 4s retained Morgan's traditional front-end treatment – that is, a folding windscreen (dropped in '56), a flat, near-vertical radiator set back from old-fashioned, stalk-mounted headlights linked by a crossbar carrying a central passing light; and unconnected wings/fenders that left the chassis unprotected, and the front suspension and underbelly naked. Two spare wheels were still carried on the tail.

Traditional it may have been, practical it was not – nor pretty, either, to modernists' eyes. In the early Fifties, remember, Morgan was more a slave to fashion than tradition, just like other manufacturers. There were also looming component supply problems to consider. The ensuing facelift – the only major one in the four-wheelers' history, discounting the failed Plus 4 Plus, of which more later – addressed these issues in 1953. It did not go smoothly.

First, Morgan 'dressed' the exposed front with a wing-connecting valance that brought the bodywork down to bumper level and gave the nose a more streamlined, filled-in appearance. It also cowled the radiator. So far, so good. However, neither the new sloping grille, topped by a badge-carrying quarter-moon trim, nor the headlights, set in ugly 'bean can' tubes, looked right. Although new regulations forced Morgan to raise the lights soon after, all for the better aesthetically, the new front was still messy. Curving the slatted grille into the radiator, deleting the quarter-moon trim and mounting the headlights in teardrop pods atop the front valance created the much-admired visage that's still with us today. It also left for historians and collectors the confusion of 'interim' models subdivided into 'low lamps' and 'high lamps'. Of the 'interim-cowled' models, thought to be 19, only a couple are said to exist in unmodified form today. The two-seater's twin-spare tail survived until 1955. After that, a single spare was laid flat to the sloping rear deck (later on, in '58, it was sloped even more).

Long before the Vanguard-engined Plus 4s had run their course, more potent Triumph TR-powered alternatives were listed, from 1953. Most buyers opted for the TR, nominally of 1,991cc for the TR2/3 and 2,138cc for the TR4/4A (unless sleeved down to under 2 litres for competition). In supplying engines to a competitor, Triumph was hardly cutting its own throat, by the way. Plus 4s were made in such small numbers they were no real threat to the mass-made TRs: in the three-year period it took Triumph to make over 58,000 TR3As, Morgan produced fewer than 1,000 Plus 4s.

Morgan four-wheelers were by now beginning to fly. Power had shot up from 34bhp to 40bhp, then to 68bhp and, with the 1,991cc TR2 engine, to 90bhp. When the 100bhp SU-fed TR3 powerplant arrived, the Plus 4 bordered on the seriously quick. Remember, the car weighed 150lb less than Triumph's stark guppy-mouthed TR2, so it had punchier acceleration, at least up to 60mph, before the Moggie's bluff aerodynamics held it back. On the other hand, the mid-mounted Moss gearbox denied Morgan the switch-operated overdrive enjoyed by the TR. Just as there was a production overlap between the

A Plus 4 with the interim radiator grille, which was introduced late in 1953 when supplies of the traditional flat radiator and free-standing headlamps were almost exhausted.

The interim radiator grille was less than an aesthetic success, and within months had been replaced by the first of the curved variety, which established the design which has been retained, with subtle improvements including a lower cowl, ever since.

Rear view of the same 1954 Plus 4. The provision of twin spare wheels on the two-seater would continue for a further year.

Vanguard and TR2 engine (most coupes continued to be made with the former), so there was between the TR2 and TR3 engines. Larger SU carburettors and a new high-port cylinder head accounted for the TR3's extra 10 horsepower.

The Motor described the wire-wheeled, £969, TR3 -engined Plus 4 it tested in 1958 as "the cheapest 100mph car made in Britain." With 0-60mph acceleration on the right side of 10 seconds, it was, said *The Motor*, "a match for competitors costing up to twice as much." It went on: "Cockpit dimensions have been increased considerably since the old days, but the car remains a carefully tailored two-seater." There was still no proper adjustment for reach, though, other than by sloping the one-piece backrest. "Once inserted, the driver finds himself unexpectedly comfortable on an air-cushion seat." As always, these pump-up cushions were separated from the road only inches beneath by wooden floorboards. Getting to the heart of the matter, *The Motor* went on: "Stiffly sprung and firmly damped, the Morgan's roadholding is outstanding, and it remains one of the very, very few vehicles (including cars built for racing) which corner with no appreciable roll." And just as well: one characteristic of sliding-pillar suspension is that the wheels remain parallel to the body sides, so appreciable roll would mean appreciable camber – and loss of grip.

The TR3 engine gave way, in the usual phased overlap, to the TR4 in 1961. Boring out to 86mm increased capacity to 2,138cc and output to 100bhp. Power went up to 104bhp with the launch in 1965 of the TR4A engine. Between them, TR3/4/4A-engined cars still account for the bulk of all Plus 4s made by Morgan – about 3,400 out of a total (to the start of 1997) of under 5,600.

With these Triumph engines, Morgans became the cars to beat in production-based sportscar racing. A leading exponent was Chris Lawrence, who won the Freddie Dixon Trophy for modified sportscars in 1959. Morgan was so

Dark brown markings on cream-coloured dials were retained through to 1962. Although the main instrument panel remained in the centre of the dashboard, a matching full-size rev-counter, when fitted, was placed as close to the driver as the speedometer.

A picture which illustrates what Morgans are all about. Three-wheelers and four-wheelers each have their devotees, who treasure their cars with a passion and a pride epitomized by the superb turnout of this Matchless-engined Supersport.

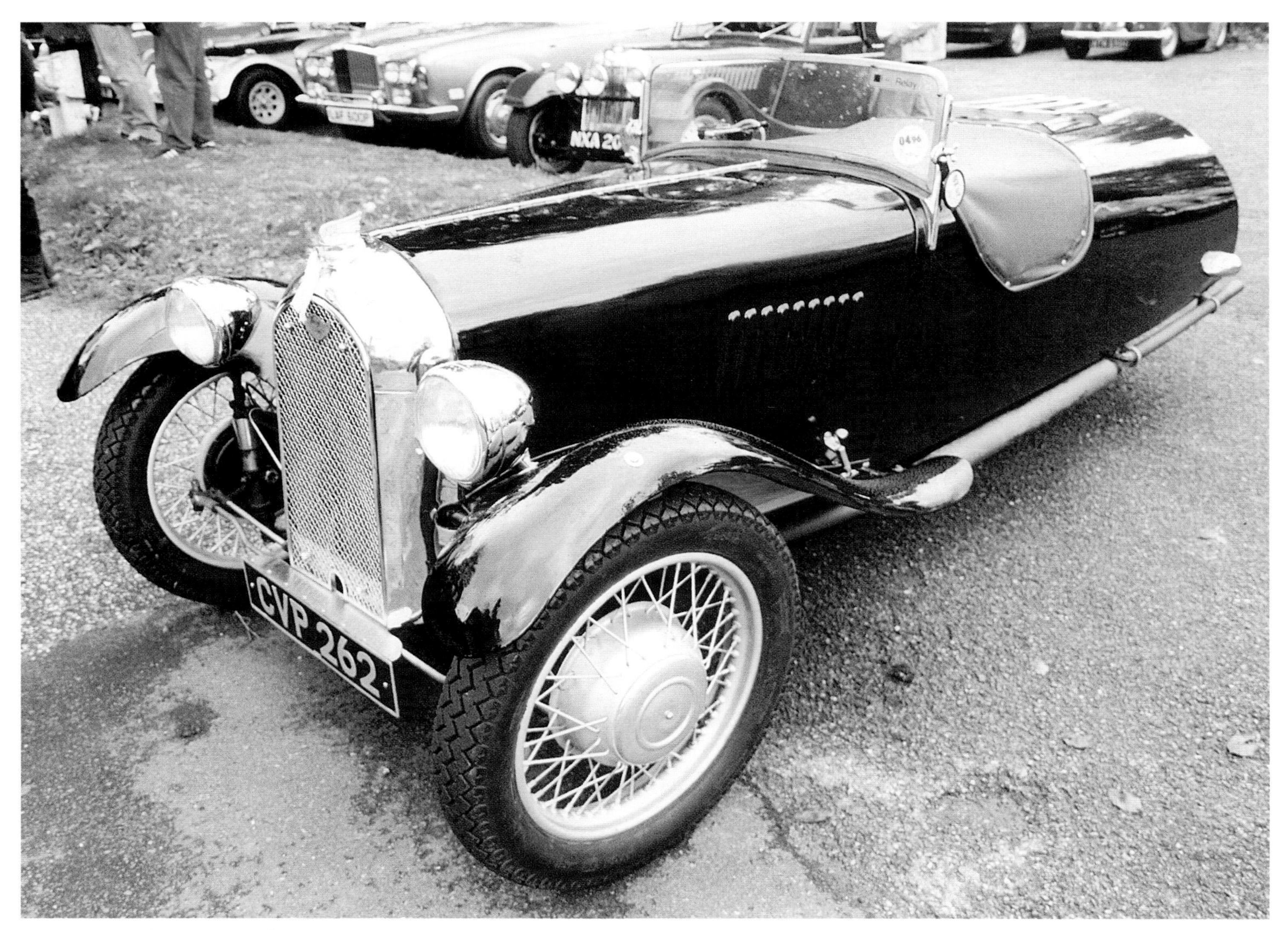

A rare F-type three-wheeler from the late Thirties wearing a radiator grille very similar to the type which appeared on the front of the first 4-4s, later redesignated 4/4s.

A left-hand-drive two-seater Plus 4 drophead coupe with neat-fitting hood, the front portion of which can be rolled back into the 'de ville' position.

An early cowled-radiator two-seater at the same location for promotional photography. Normally, indicator lights would be recessed into the wing aprons beneath the headlamps.

impressed it commissioned from Lawrence a limited supply of specially tweaked TR3/TR4 engines.

Thus was born, in 1961, the Plus 4 Super Sport – today one of the most prized of all Morgans, yet still comparatively affordable. Lightweight aluminium bodywork, optional on other models, was standard. So were 72-spoke wire wheels; few sportsters look better on 'wires'. Nothing drastic was done to the engine, which drew its air from an offside bonnet blister, riveted on to begin with, later (when the early high-nosed cars gave way to low-line ones), blended in. It was the cumulative effect of many tweaks – polished ports, a gasflowed head, a higher compression ratio, twin Weber carbs, a four-branch exhaust, special cams and Brabham-balanced internals – that bumped up power to a safe 115bhp (TR3) or 125bhp (TR4) and top speed to 120-125mph, though wilder tuning yielded more power and speed. In standard trim, the Super Sport would accelerate as quickly – 0-60mph in 7.8 seconds – as the first of the Plus 8s that temporarily curtailed Plus 4 production several years later.

In racing trim, the TR engine delivered over 150bhp, and Lawrence made good use of it. Driving with Richard

Shepherd-Barron, he amazed the racing fraternity with a class win at Le Mans in 1962 (see Chapter 8). Morgan made 101 Super Sports before the Plus 8 effectively ended its production. As a cut-price alternative to the semi-exotic Super Sports, Morgan also produced the steel-bodied Competition Plus 4, with mildly tweaked TR4A engine and

One of the rare Plus 4 four-seater drophead coupes, sometimes irreverently referred to as 'Smog Mogs'. Only 51 of them were produced between 1951 and 1956.

Unusually for a Morgan, the spare wheel of the four-seater coupe is hidden from view in its own locker, leaving very little space for luggage. Despite quite a large rear window, blind spots in the quarters of the hood must make reversing round corners a tricky manoeuvre.

low-line bodywork like that of the Le Mans car. Only 42 of them were made, most of them being exported. From 1966, all Plus 4s were given low-line bodywork, incidentally.

Besides the high-profile Super Sports and Competition, there were two other variations of the Plus 4 theme before the series was temporarily mothballed in 1968 – the year of the Plus 8 covered in the next chapter. One of these was the little known four-seater version of the drophead coupe, nicknamed the Snob Mog. Only 51 were produced, most with a Vanguard engine (though TR engines powered some exported cars). The other was the Morgan that everyone does remember, if only for the wrong reason. This was the Plus 4 Plus – the bubble-topped, glassfibre-bodied, all-enveloping streamliner commissioned in 1963 from EB Plastics as an alternative to the traditionally-styled car.

"I didn't have the nerve to discontinue the old one," Peter Morgan told *Car and Driver*. "And just as well. People came up to me at the show and said: 'Well, I don't know how much I like that one. Looks a bit like a Lotus at the front

From any angle, this 4/4 two-seater looks like a prize-winning presentation. The 12-hole wheels and vertically slatted radiator grille were fitted to most examples of this model from 1938 onwards.

The 4/4 disappeared from the Morgan catalogue in 1950 when the supply of Standard engines dried up, but it was back again, under Ford power, in 1955 and has been part of the Morgan range ever since. This gleaming example is a 1970 two-seater.

Enthusiastic Morgan owners travelled long distances to attend Mog 95. The owner of this Plus 4 two-seater drophead coupe had journeyed all the way from Spain.

and a Jaguar at the back.' So I scrapped it. In the mid-Sixties, the old car started selling again, and we've never looked back since."

Peter Morgan offered an explanation for the Plus 4 Plus' failure. "I think it's a reaction to the computer-designed car...the Morgan looks like a car the way people imagine a car should look...a child can see it very clearly: the wheels are here, the engine goes here, the spare wheel goes here, and we sit here and drive with this..."

Look through half-closed eyes and you *can* see elements in the Plus 4 Plus of the Jaguar XK150 and original Lotus Elite, just as there was a strong hint of Frua's AC 427 in an alternative open-cockpit design submitted by EB Plastics. With wind-up windows, chromed window surrounds, a separate boot, even outside door handles, the Plus 4 Plus was like no other Morgan, before or since. Under the skin, though, nothing had changed other than the addition of some sheet-steel reinforcement to minimize the risk of bodywork cracks: Morgan knew only too well how much its chassis flexed.

At £1,275 7s 1d, the TR-engined Plus 4 Plus was at the time the most expensive Morgan ever made. However, it was not the price so much as the car's style and character that alienated buyers. Striking though it was – quick and well-finished, too – Morganists wanted the real thing, not a pseudo modernist. Only 26 were sold in the 1963-67 period before Morgan conceded its error and terminated production. Perversely, the Plus 4 Plus is today one of Morgan's most cherished models, not least because of its rarity.

Morgan learnt several lessons from the Plus 4 Plus. The first was to be very wary of 'advice' from outsiders who perhaps didn't understand the Morgan ethos and its unique appeal. (Much later, in 1990, Morgan's directors backed their own instincts and rejected the televised recommendations of the BBC's troubleshooter, Sir John Harvey-Jones, to modernize and expand.)

The second was that Morgan's future was rooted in the past. It confirmed that the basic appeal of its cars was their nostalgic lineage. Morgans were not replicas, but living

An interesting transplant. Beneath the bonnet of this Plus 4 four-seater nestles a Rover 2000 engine.

vintage sportscars that provided an exhilarating antidote to modern motoring. Sleek streamlining didn't enter the equation, even though it gave the Plus 4 Plus a top speed of over 110mph. In that it focused attention on Morgan's real virtues, the bubble-topped failure did the company no serious harm, and perhaps a great deal of good.

One can only speculate about Morgan's fate had the Plus 4 Plus been built the other way about – with visually unchanged sheet metal and modern underpinnings. A torsionally stiff Morgan with wishbone front suspension and a decent multi-link rear end? Think about it. Morgan certainly is (see Chapter 8).

With the demise of the TR-engined cars, the Plus 4 series faded from the scene while the Plus 8 stole the show. However, 17 years on, in 1985, the series was revived under curious circumstances. As recounted in the last chapter, Morgan had replaced the moribund Cortina pushrod engine in the ongoing 4/4 with the 1.6-litre Fiat Twin-Cam before Ford, whose association with Morgan went back to the three-wheelers of the Thirties, came up with a 1.6 CVH alternative. Morgan did not so much ditch the Fiat engine as change it for another – a 2-litre job, complete with five-speed gearbox, that inspired the Plus 4's resurrection. Said *Motor*: "When considering alternatives such as the Saab 2-litre and small BMW sixes, Morgan decided that the Italian engine was the best bet."

Morgan's new mid-range model was based on the 4/4's chassis (wheelbase still 96 inches), rather than the Plus 8's wider, longer one (adopted in the interests of rationalization in 1991), though cut-down wings from the Plus 8 cradled the newcomer's 6x15 Cobra wire wheels, shod with generous 195/60 rubber. Low-profile radials on 'wires'? It was certainly an unusual combination. Evolutionary changes to the suspension included softer (but still hard) front springs and a more supple five-leaf arrangement (formerly six) at the back. Cam Gears' obsolete worm-and-nut steering had given way to French Gemmer recirculating-ball. At 12.5 gallons, the Plus 8-like fuel tank was usefully bigger than the 4/4's.

Although quite an old engine in 1985, the Fiat Twin-Cam

The availability of the 2,088cc Standard Vanguard engine led to the creation of the Plus 4 in 1950. This particular car earned considerable fame as the one used through 1951-2 by Bill Boddy, the founder editor of *Motor Sport* magazine.

A 1966 Plus 4 Plus, one of only 26 examples of this glassfibre-bodied coupe to be produced, due to lack of demand. Its appearance, however, shocked Morgan enthusiasts into a renewed interest in the traditionally styled models, sales of which had been slow.

had high-tech accoutrements, including Bosch LE-Jetronic fuel injection (except for nine Weber-fed cars) and Marelli Digiplex ignition, much to the horror of do-it-yourself Morganists accustomed to tweaking their own engines. With quoted outputs of 122bhp and 127lb ft of torque, there was no shortage of Latin *brio*.

But *Motor* had reservations about the £10,901 Fiat-engined Plus 4 it tested in '85: "The engine doesn't feel particularly sporting and its marked mechanical harshness above 5,000rpm deters frequent use of the full range...the best technique is to use the engine's excellent torque, keeping the engine revs between 3,000 and 4,500rpm...the SEAT-built gearbox has a stubby lever and short, precise action but it is never light or notably rapid." *Autosport* was more enthusiastic about the powertrain: "One of the best things about this Plus 4 is the engine/gearbox combination."

Although competitive with the opposition on acceleration (0-60mph in under 9 seconds), brick-like aerodynamics as always restricted top speed (112mph). Mind you, ton-up

The name John H Sheally II is synonymous with highly polished chrome and paintwork. The fastidious American owner and restorer of Morgans was responsible for rebuilding this dazzling Plus 4 Super Sports, which now resides in Holland.

The Plus 4 Super Sports, identifiable by the slim air intake on the right side of the bonnet, was introduced in 1961 and benefited from a Chris Lawrence-modified and balanced Triumph TR engine, which increased peak power to approximately 125bhp.

Another view of a Plus 4 Super Sports, this example being from 1967 production and showing the improved rear vision through the erected hood, and the sloping tail, which had been introduced in 1955.

Another interesting engine transplant, this time the Daimler SP250's V8 power unit, which has been inserted into this Plus 4 Super Sport, pictured at Mog 86.

wind noise discouraged high speeds in this and every other Morgan. *Motor* liked the chassis changes. Its testers reckoned the steering, now free from frictional stiffness, was even better than the Plus 8's rack-and-pinion setup. Just to confuse the issue, *Autosport* contradicted: "The steering...feels heavy and sticky, so the initial impression of the handling is one of obdurate understeer." Still, both magazines agreed about the roadholding. Said *Motor*: "For cornering ability, the Morgan is as enjoyable as it is surprisingly good."

Over a two-year period, from 1985 to '87, Morgan made 122 Fiat-engined Plus 4s before supplies dried up – much to the relief of those Bulldog Brit enthusiasts who found it hard to accept an alien engine under the bonnet of a quintessentially English car. Morgan has not ventured abroad on a shopping spree since. Instead, it turned to its own back yard, to Rover's potent M16 – an in-house engine (rather than a Honda disguised by Rover nameplates) of modern twin-cam, 16-valve design from the front-drive 800. Sherpa van parts were employed to marry the 136bhp

The Morgan Motor Company's famous number-plate, which over a long period has adorned a variety of cars, is attached here to a Plus 8 for a publicity photograph against an appropriate Worcestershire backdrop.

Morgans and motorsport have always been synonymous, and here Fred Scatley's camera has caught Rob Wells and Patrick Lund battling hard during a race at a Bentley Drivers' Club meeting at Silverstone in 1993.

M16, converted from a transverse installation to a longitudinal one, to a Plus 8 (Rover) gearbox. I well remember attending the 1988 press launch of this car at Brands Hatch, not least because it was the first proper launch (instigated by Charles Morgan), embracing test driving, speeches and a good nosh, that Morgan had ever staged. Prices (excluding extras) ran from £13,500 for the two-seater to £14,500 for the four-seater.

When Rover's M16 gave way to the cat-cleaned T16 – a more modern variation of the twin-cam 16-valve theme – Morgan was obliged to follow suit. The reduction in horsepower, down to 134bhp from 138, was hardly noticeable, especially as torque was greater. Morgan took this engine-swap, which occurred late in 1992, to upgrade the Plus 4's chassis, hitherto shared with the 4/4, to that of the Plus 8. Subsequently, Rover – and therefore Morgan – evolved the T16's 'black box' Motorola ECU (electronic control unit) from MEMS 1.6 to MEMS 1.9 to include anti-theft immobilizer circuitry. Like the 4/4, the Plus 4 got

One of the three SLR (Sprinzel Lawrence Racing) Morgans built from 1963 with more aerodynamic aluminium bodywork over a conventional Plus 4 chassis for use in Group 4 GT racing, in which they achieved considerable success.

More than three decades later, this SLR Morgan, equipped with Nineties-style tyres and wheels, is still enjoying an active competition career, but carrying number-plates as a reminder that this is also still a road-going car.

The Plus 4 Plus was considered by Morgan's management to be a necessary model to take the company forward in the rapidly changing car market of the mid-Sixties. Regular customers had other ideas. Only 26 would be made in just over two years while demand for the traditional models increased.

upgraded brakes – a servo-assisted Lockheed disc-drum set-up with four-piston front calipers and self-adjusting rear drums – in 1993.

March 1994 is a significant date in the Plus 4's evolution (and the Plus 8's too, for that matter). It was then that the old five-speed gearbox gave way to Rover's new R380, as used in the Range Rover. It came with several advantages. For a start, the box was quieter and stronger. First and reverse, formerly both gated left and up, were moved to opposite ends of the movement. Lightening the lever also improved shift quality. As the new gearbox was an inch longer than the previous one, Morgan had to create an underslung crossmember to carry it.

Come the summer of 1997, all Morgans had been engineered – with more rearward seat travel, longer doors and a deeper dash (as described in Chapter 4) – to carry airbags.

Here, chronologically, is a summary of the Plus 4's development from 1950 to 1996:

October 1950
Prototype Plus 4 built. Makes Earls Court debut powered by 68bhp Standard Vanguard engine. Car goes on sale the following year with engine-lubricated front suspension. First left-hand-drive car soon follows.

September 1951
First of the Plus 4 four-seater drophead coupes; only 51 produced, the last in February '56. Also, one-off four-seater saloon built to special order. Optional Smiths fug-stirrer heater offered the following year.

October 1953
First new-look 'interim radiator' Plus 4. Also, first of the 90bhp, TR2-engined cars. TR2 engine standard by spring of '54.

June 1954
New-look 'high-cowl' styling replaces old 'flat-rad' and 'interim' models. The following year, a single spare wheel is recessed into deck of new sloping tail after trunnion spring mounts give way to shackles.

February 1956
First of the 100bhp TR3-engined Plus 4s. TR3 engine standard by March '56, though last Vanguard-engined car is not made until summer of '58.
October 1958
Minor styling changes include revised rear end (all models), still with single recessed spare wheel and widened body. Fuel tank enlarged, running boards narrowed, amber flashers standardized. Adjustable seats, wider than before, introduced. New Girling brakes from December.
August 1959
Girling disc brakes offered as an option (front only) with wire wheels. Steel wheels, too, from chassis 4368. Front discs standard by November 1960.
September 1960
Revised dash; cream facia knobs give way to black toggle switches.
February 1961
First Plus 4 Super Sports. Over next seven years, 104 built, the last in May '68.
June 1962
2.2-litre TR4 engine standard after being offered as an option. TR3-engined models continue until stock exhausted. White-on-black dials introduced in another facia change.
November 1964
Ill-fated Plus 4 Plus introduced. Only 26 made before model axed as commercial failure.
October 1965
All cars now with negative-earth electrics. Competition model introduced with low-line body of Super Sports; made until November '66.
December 1966
Low-line body adopted for two-seaters.
May 1967
TR4A engine with 2,138cc and 104bhp.
September 1969
Last TR-engined Plus 4 (a drophead coupe) made. Production of two-seaters and four-seaters had ended previous year.
October 1985
Plus 4 reintroduced with 122bhp 2-litre Fiat Twin-Cam engine and five-speed gearbox. Slow-selling model – only 122 made in three-year period.
1986
As for other models, durability improved by various measures, including galvanization, powder coating, Cuprinol dipping, wings-off painting and so on. Nuts, bolts and fasteners later passivated or zinc-plated.
May 1988
Rover M16 2-litre 16-valve twin-cam introduced. Mated to same five-speed Rover gearbox as in Plus 8, using parts from Sherpa van. Power (138bhp) and torque (131lb ft) both greater than displaced Fiat's. Battery relocated to rear in '89.
December 1991
To rationalize production, Plus 4 now made on Plus 8's wider chassis, increasing track. Rack-and-pinion steering offered as option to Gemmer recirculating-ball.
1992
M16 engine displaced by more modern T16 in line with Rover's evolutionary product programme. Plus 4 chassis same as that of Plus 8 from here on.
1993
Servo-assisted Lockheed disc/drum brakes, with four-pot front calipers and self-adjusting rear drums, replace unassisted Girling ones.
March 1994
New R380 Range Rover gearbox – stronger, quieter and smoother-shifting – displaces old Rover five-speeder.
Jan/Feb 1996
Engine's electronic black-box controller gets anti-theft circuitry. Traditional GKN Salisbury back axle displaced by new BTR one sourced from Australia.
Summer 1997
To facilitate airbag installation, all cars get extended doors, more seat travel, a deeper dash and dished steering wheel. Stainless steel exhaust system reduces noise. By now, all Morgans are being sprayed in new dust-free paintshop.

CHAPTER 6

The blistering Plus 8

A Morgan with Anglo-American appeal

With the demise in sight of Triumph's old 2.2-litre TR4A 'four', last of the Vanguard derivatives, Morgan was looking for new engines by the mid-Sixties. The pushrod straight-six that Triumph had lined up for the TR5 had the right credentials, not to say a deliciously addictive exhaust rasp. Unfortunately, it was too bulky and long. Despite the span of the Morgan's bonnet, space beneath it was limited by footwells that extended well forward, pushing the engine towards the nose.

Morgan's other main engine supplier, Ford, had a compact V6 of the right output, which was being used in the Zodiac and Capri (in Reliant's Scimitar, too), but it was a hefty all-iron lump that sat too high under the bonnet. Potent though it was, the Lotus Cortina's twin-cam was deemed too frenetic for the job. Besides, once again, taking the Ford route would have made the company totally dependent on one engine supplier, and Morgan was wary of that.

Triumph's new 3-litre V8, destined for the ill-fated Stag, was just over the horizon, and was informally offered to Morgan by engineering chief Harry Webster. But Morgan turned it down – and just as well in view of its subsequent teething troubles. Morgan could have ill afforded the crippling warranty claims that would probably have come with this troublesome motor.

Talks between Peter Morgan and Rover's Peter Wilks had touched on a possible takeover bid that was firmly rejected. More fruitfully, the meeting put the lightweight, all-aluminium V8 that Rover had bought, lock, stock and tooling, from General Motors firmly in the frame at Malvern Link. If Rover couldn't have Morgan, to be associated with it through an engine was the next best thing. Rover had not supplied engines to anyone else before, but there was always a first time.

GM had ditched its lovely alloy V8 only because of the development of new thin-wall casting methods that made iron blocks cheaper to make and almost as light. There was nothing innately wrong with the engine. Nothing at all: 750,000 were running around North America in Buick Specials, Pontiac Tempests and Oldsmobile Cutlass F85s. What GM now saw as surplus to requirements, Rover's managing director, William Martin-Hurst, saw as treasure trove. It was a formidable coup to transfer production to Britain under Rover's nameplate.

By 1967, the Anglo-American V8 was powering the last of the 'auntie' Rovers, the P5 3.5. The following year, it was shoehorned under the bonnet of the advanced P6 3500, mated to an automatic gearbox. But why stop there? The engine had huge potential for other applications, as Morgan, TVR, Land Rover and others were soon to demonstrate.

Rover's new engine was not the first V8 to power a Mog. Just before the War, in the late Thirties, Morgan had experimented with Ford's ubiquitous side-valve V8, like that subsequently used in the big Pilot saloon. The experimental car went like a rocket. The trouble was, it didn't stop very well. It would also have been taxed very heavily under the iniquitous RAC horsepower rating system still in operation. Morgan abandoned the project.

Race engineer Maurice Owen, late of the wound-up Laystall Formula One racing team, joined Morgan in 1966 to build in total secrecy the second V8 prototype, this one with a Rover engine, primarily to see if it worked. It was a hands-on, fix-it-as-we-go job, conducted in a small, concrete-floored, brick-built outbuilding, grandiosely known as the research and development department.

The starting point was a strengthened Plus 4 chassis – what else? – and a Buick V8 that by sheer coincidence Owen had bought at auction, as a job lot of three, when Gordon-Keeble's assets were sold. Owen told *Classic and Sports Car* magazine in 1997 that one engine went into a Cooper hillclimb car and a second into a Rover 2000TC (before Rover got around to doing the same thing). For the third, Owen had tried to buy an engineless Plus 4 chassis. Instead, he was asked to join Morgan and do the job in-house – officially.

There were, of course, problems. Although the 3.5-litre alloy V8 was no heavier than the 2.2-litre all-iron 'four' it displaced, it was considerably wider and wouldn't fit without surgery. That the existing steering gear had to be modified, using a collapsible AC Delco steering column, was no bad thing as Morgan had to take on board modern safety accoutrements if it was the sell in the US. The engine's tall SU carbs, swapped from Holleys, were accommodated on the prototype by bonnet bulges. Hammering a dent into the cylindrical air-cleaner, so that it would fit under the bonnet hinge, became a part of Morgan folklore. So is the story of the restorer who 'undented' the box, only to find it then wouldn't fit.

There was no room for an engine-driven fan, either, so in its place went a thermostatically-controlled Woods-Jeffreys electric one, foreshadowing what was to become common practice elsewhere. The Plus 8 was also the first Morgan to get an alternator to charge its 12-volt rear-mounted battery. Extra equipment – foglamps, hazard lights, triple wipers and so on – meant a more complex electrical system, too.

An early production version of the Plus 8, wearing one of the Morgan Motor Company's familiar number-plates, parked outside the factory in Pickersleigh Road.

The Plus 8 became progressively wider as the years went by, this being the early narrow-body version, identifiable in this view by the proximity of the headlamps to the radiator cowl.

The same car from the rear, with just a pair of overriders, although a full-width bumper was offered as an optional extra. Note the black instrument dials and the smaller rev-counter than on earlier Plus 4s.

Retention of the four-speed Moss gearbox, which didn't cramp the footwells like later Rover boxes did, meant modifications to the magnesium bellhousing, flywheel, Borg and Beck clutch and primary driveshaft. It was a tribute to the gearbox's strength that it could take so much more torque without modification. Refinement was not its forte, but toughness certainly was. Mechanical clutch actuation gave way to hydraulic, and Owen made detail changes to the suspension, still lubed on a daily basis (or every 200 miles) up front by a cockpit pedal that squirted engine oil under pressure to the uprights. He beefed-up the stub axles, so they were stronger than those of the four-cylinder cars. With so much power on tap, there would be no skimping on brakes. He used a Girling disc/drum system (soon upgraded to dual circuits) with strengthened calipers and servo-assistance – another first for Morgan. Owen increased spring travel a little, to around 4.5 inches, realigned the rear leaf springs and introduced a limited-slip diff in an attempt to curb axle tramp.

In the end, it was not so much technical hitches as political and bureaucratic ones that delayed production. Morgan had little trouble obtaining GM's permission to use its old engine. It was the 1967 merger of troubled Rover and British Leyland (which already embraced Triumph) that caused the biggest hiccup. Morgan's needs were hardly a priority issue.

The wire-wheeled prototype, OUY 200E (chassis number R7000), looking much like a Plus 4 with its narrow hips, first ran in February '67, Maurice Owen at the wheel in a heartening night-time debut that exceeded expectations (the car remained with Morgan for some years before it went to an American enthusiast). It was not until the following year, though, after permission from GM to use the engine had been granted, that Plus 8 two-seaters finally started to roll

An early Plus 8, in this instance a 1969 model, amongst a Morgan gathering at Beaulieu. On cars built up to 1972 the Rover V8 engine was coupled to a four-speed Moss gearbox.

from Pickersleigh Road at the rate of 15 or so a month. Morgan collected the engines from Solihull itself. Only after BL luminaries Harry Webster and George Turnbull had driven the prototype – and been suitably impressed – was engine supply assured. Two cars were displayed at the '68 London motor show, and a handful of Plus 8s were delivered by the end of the year, all with steel bodies and very basic Ambla trim.

Before production was seriously under way, Morgan extended the Plus 8's chassis in every direction. Its width increased by a couple of inches, from 47 to 49, its wheelbase (and therefore its length) by a similar amount, creating as a by-product slightly more room than in the Plus 4 that had spawned the new model. Over the years, the body got wider still, all the better to accommodate additional rubber. Fatter tyres meant more grip – and the Plus 8 needed all it could get. The downside is that without power-assistance (never fitted to a Morgan), it also meant more steering effort: the first Plus 8 was a pig to park, and later ones were even worse.

Another early Plus 8, with some visible extras including a full cockpit tonneau, a useful luggage grid over the spare wheel and the full-width chromed rear bumper. From 1971 these upturned rear lights were replaced by vertically faced units on tubular mounts.

A Plus 8 with the rough-cast alloy wheels which were fitted through to 1976. The twin spotlamps, closely flanking the radiator grille, were part of the Plus 8's standard equipment from the start of production.

A Plus 8 running on non-standard wide-rim wheels and equipped with the tubular-mounted rear/stop/indicator lights introduced in 1971. Originally these were supplied chromed, but many owners have subsequently had them painted body-colour, in line with cars built from 1987 onwards.

Unprotected wooden floorboards had given way to painted steel (Morgan later reverted to wood), in a bid to stiffen the chassis, and an increase in fuel tankage to 14 gallons (64 litres) reflected the V8 engine's extra thirst. It was served by two fillers, one each side on the sloping rump, that left no room for a Morgan badge. Three screen wipers, *a la* Jaguar, were also fitted (as they were much later on, in the Seventies, to other Morgans).

Although the style had not changed – Morgan was not going to repeat the Plus 4 Plus mistake – the Plus 8 was a little bigger and more purposeful in its stance than any previous model. Its wings (fenders) were extended to accommodate 5.5in-rim rough-cast aluminium alloy wheels (specially made for Morgan by Robinson) and Dunlop SP Sport 185x15VR tyres with Aquajet treads. The Plus 8 was not at the time, or for many years after, offered with wire wheels, these being considered of inadequate strength for a high-performance car. Today, though, you can specify Indian-made 'wires' that have passed a rigorous MIRA strength test.

In place of Morgan's traditional air-cushion seats, linked by a common backrest, Owen drew on his competition background and recommended slim Restall buckets, with

A 1973 Plus 8, with the Rover four-speed gearbox, climbing the old test hill at Brooklands. Later cars had five-speed 'boxes.

Ambla vinyl or leather trim, for the Plus 8. They were much more supportive in their embrace than previous Morgan seats. What's more, they were adjustable. Decent support was now vital because the big V8 elevated the lightweight (17cwt) Morgan into another performance dimension.

Further concessions to safety included flush rocker switches on a facia that owed more to symmetry than ergonomics; the white-on-black speedometer was difficult to see over on the passenger's side, and the rev-counter (of different size and design) was partially masked by the wheel rim on the right. A single Ford Corsair column stalk worked the horn, dip and indicators. Small specialists like Morgan always have trouble with the dash – a major design and tooling expense for mainstream manufacturers who would spend more on designing a facia than Morgan would in creating a new car. The Plus 8 is said to have been developed for under £15,000.

There was nothing particularly special technically about the 90-degree engine other than its lightweight construction, which suited Morgan just fine. It was fashionably 'over-square', with an 88.90mm bore and 71.12mm stroke – a far cry from the long-stroke, tax-beating prewar 4/4s – giving a swept volume of 3,528cc. The pistons ran in steel liners. Two HS6 SU carburettors fed the light-alloy, crossflow cylinder heads, each carrying eight valves worked through hydraulic tappets by pushrods running from a single chain-driven central camshaft. The cast-iron crank was carried in five bearings.

As the American saying goes, there's no substitute for cubic inches. Despite the engine's modest state of tune, power leapt from the Plus 4's 100-115bhp to 168bhp net and torque to 210lb ft. Where the Plus 4 had been merely quick, the Plus 8 was supercar quick. Top speed at over 120mph was nothing special, for all the usual bluff-fronted reasons, but the car had blistering acceleration that no rival could match. Cars of similar price, like the MGC, Lotus Elan and Triumph TR5, were left standing by the mighty Mog, while the E-type Jaguar cost considerably more.

A few home truths from *Road & Track*, testing a personal import to the US in 1969 (before the Plus 8 was officially

The earlier dashboard, with the two major dials mounted in a central panel, continued in production through to 1976. Triple windscreen wipers were a standard fitments from the start of Plus 8 production.

sanctioned for sale in America), did the car's reputation no great harm: "It rides hard, steers hard, isn't put together very well, leaks rain, has a token heater, a cranky gearbox...but a Morgan is nostalgia on wheels – genuine nostalgia, not plastic nostalgia." *R & T* had trouble with the gearbox, still bereft of synchromesh on first, and not over-endowed with it on second, third or top, either. "The vintage gearbox nearly spoils it," wrote the magazine's testers, who liked the engine, roadholding, cockpit and new Restall bucket seats.

This privately-owned customer car was not as quick as Morgan's press demonstrator, MMC 11 – the second prototype, chassis number R7001, that incorporated all the production modifications except the triple wipers – tested by *Motor*. I remember it well. How could you forget a car that kicked and bucked, went like a rocket, demanded your full attention when extended and assailed the ears, never mind stirred the soul? For all its many faults, I loved that first Morgan Plus 8. It still exists (though not with its original chassis) and as most devotees of the marque will

Peter Morgan pauses during a test run around the hills near Malvern with an early Plus 8.

This excellent Vic Berris cutaway drawing featured in the model description of the Plus 8 in *Autocar* on the occasion of the car's introduction in September 1968.

know, the MMC 11 registration is Morgan's most famous.

Said *Motor* in its test of MMC 11: "The Rover engine in its Morgan installation is a remarkably fine marriage...the lightweight V8 is extremely smooth and delivers its power in such a subdued and unfussy manner that really high speeds become deceptively effortless." I was never much bothered by the Moss box's weak synchromesh, which besides its own inertia, had to overcome that of the primary driveshaft as well. The simple answer was to double-declutch, especially from first to second, which rewarded deftness with smooth, snick-free shifts – and also exercised your left thigh on the heavy clutch.

Despite the stiffened chassis, scuttle shake was still part and parcel of Morgan masochism. Stiff steering and a gut-jarring ride, too. Said *Motor*: "Bad roads generate loud bonks from the front suspension and a very pitchy ride. Bumps on corners can throw the car off line, too." It also spoke of rattles, wind buffet and hood-up noise. There may have been mighty progress under the bonnet, but little else had changed, thank goodness. Three decades on from the original '67 prototype, the Plus 8 has not changed fundamentally, though there have been so many detail improvements that the next chapter is devoted to the car's steady development.

CHAPTER 7

The Plus 8 matures

Lower, wider and more powerful

The Plus 8 is the longest-running model with the same engine in Morgan's history. It is also the most numerous of the four-wheelers (it overtook the second-placed Ford Kent-engined 4/4 some time in 1991), and the only one to be built exclusively as a two-seater, dismissing the four-place one-off special made for agent Eric White.

The series started with chassis number R (for Rover) 7000 (the first prototype) and had reached at the end of 1996 in a straight – and by Morgan standards, relatively uncomplicated – run around R11,400. For every year after that, add 200 if you want a rough guide to Plus 8 production numbers. To put these figures into perspective, Ford sells twice as many Escorts in Britain every month as Morgan has sold Plus 8s worldwide in nearly 30 years.

Although the latest cars are bigger and heavier than the early ones (more about this later) they look much the same until you delve beneath the skin. Under engineer Maurice Owen, who went on to become Morgan's long-serving technical director and a board member, evolutionary development has been a continuous progress spanning over a quarter of a century. And there's more to come yet.

Many changes made to the Plus 8 have been dictated by regulations affecting safety, noise and exhaust emissions. "It's been an absolutely daunting business," explains Peter Morgan. "We've kept up with the regulations by tackling them gradually." Other changes have been caused by component supply problems, to which Morgan has always been particularly susceptible. In its quest to build a better and more durable product, Morgan has also introduced many quality improvements in recent years.

Today's Morgans are much better protected from rust, wood rot, paint cracks and wear than they used to be. Powder coating and chassis galvanization, optional at first in the late Eighties, but standard now on the Plus 8, has extended life to respectable levels. Ditto Cuprinol treatment – now standard procedure – of the body's ash frame, and (optional) wax sealing of the whole underside. And modern flexible paint gives a longer-lasting finish.

One of the first evolutionary developments was to offer left-hand drive to satisfy strong overseas demand from the US and Europe. Germany, land of smooth roads, where sliding-pillar suspension works best, was to become Morgan's top export market, justifying two agents where most other countries had one. Morgan's return to America in the early Seventies, though, was short-lived. When Rover, its engine supplier, pulled out, Morgan had no option but to follow it to the door. No quarter was given by the authorities: to satisfy US emissions laws, Morgan would have to conduct its own 50,000-mile emissions test, just like everyone else. Spread over a few dozen sales, rather than several thousand, the cost and engineering effort could not be justified, so Morgan withdrew from the US and turned instead to less demanding markets.

That Morgans were on sale in the US again four years later is down to the extraordinary efforts of the enterprising Californian distributor Bill Fink, of Isis Imports. He beat the bureaucrats by exploiting a loophole in the regs that allowed the sale of LPG gas-powered Morgans, some of

A Plus 8 with a set of 14-inch eight-hole Milrace wheels as supplied for five years from 1977, and the aluminium bumpers, which replaced the chromed variety at the same time.

them turbocharged, provided they were fitted with various safety equipment – inertia-reel belts, a bigger screen, even 5mph impact bumpers (using VW hydraulic pistons attached to reinforced Morgan bumpers) and side-intrusion bars (the door pillars were reinforced to take them).

Don't scoff at the idea of a propane-powered Morgan. For a start, LPG gas burns very cleanly, with minimal oil contamination, so it's good for engine longevity. With turbocharging, it can also provide decent performance. The US monthly, *Road & Track*, wrote of one of Fink's 220bhp LPG turbo conversions in 1980: "A twist of the key and the V8 throbs to life. Snick the stubby lever into first...and you're burbling away, looking down that long, louvered bonnet at one of the classic views in motoring. Stomp on the throttle...and gobs of torque at the bottom end turn into raspy turbo power as the revs build...you can go from 0 to 60mph in 6.8 seconds, and turn the standing quarter in 15.1 at 93.5mph." Sounds like any other Plus 8, wouldn't you say?

With only a little tongue in cheek, *R & T* described the chassis of less than state-of-the-art rigidity as half the suspension. "The tyres are the other half...think of the springs as merely something to grease occasionally."

Because of the Plus 8's huge potential as a track racer, Morgan introduced the option of lightweight aluminium bodywork late in 1969 (though steel was retained for the bulkhead to preserve strength). These ally-bodied cars were later to spawn, in 1975, a select 19-off run of wide-wheeled, wide-tyred Sports Lightweights, which enabled owners to compete in production sportscar events barred to them with the previous non-standard aluminium skin. The weight saving of nearly a hundredweight (50kg) benefited the performance and economy of the production road cars, too, particularly when power was decreased by emissions control. As it also reduced, if not totally resolved, the problem of rust, it's not surprising that about half of all the Plus 8s made to date have had aluminium bodywork, even though it is more vulnerable to denting and damage.

Because of Custom and Use regulations, the rear lights and indicators, hitherto flushed to the sloping tail (and therefore pointing up), were mounted on horizontal chromed-plastic tubes in 1971 so they presented a vertical face to following traffic. Other models got these better-looking lights, body-coloured from '87, as well, of course. By the early Seventies, the remorseless demand for greater safety was seriously occupying Morgan minds.

Variations on the Plus 8 theme were not: although an experimental automatic drophead (1971) and four-seater (1972) were made, neither went into production.

Morgan had known for some time that the days of the strong but primitive four-speed Moss gearbox were numbered when Jaguar, its main customer, progressed to an in-house design. The logical replacement for Morgan was Rover's own newly-developed four-speeder. So, after 494 Moss-boxed Plus 8s had been built, Morgan ditched its mid-mounted gearbox and bowed to convention. Attaching the 'box to the back of the engine, as Rover did in the 3500S, meant chassis crossmember changes by Maurice Owen and the use, for the first time, of a remote-control linkage to bring the lever back to a handy position. It also meant clash-free shifting, without double-declutching. The downside was reduced room in the footwells. Even today, there's nowhere comfortably to rest your off-clutch foot.

The first Rover-boxed car was R7475, but don't be misled by that as there are Moss-boxed cars bearing higher numbers. Rarely in Morgan's production history has there

Rear view of the same car, offering a reminder that number-plates can be unreliable dating evidence. This plate suggests a 1973/74 model, but the dashboard points to 1977.

This Morgan Plus 8 demonstration model looked particularly smart in a metallic grey paint finish with highly polished alloy wheels.

From 1982, these 15-hole cast magnesium wheels became the standard fitment for the Plus 8. Note the offset number-plate, no doubt to assist airflow into the area of the oil cooler and sump of this competition-prepared car.

been an unequivocal cut-off point between specification changes. They tend to be blurred, spread randomly over a period that can run into years. If in doubt about an old car's original spec, check it out with the factory. Armed with a chassis number, it can determine from hand-written ledgers exactly what the car was and had in the way of equipment when it was built. Keep in mind, however, that over the years a Morgan's specification is quite likely to change. Things tend to get upgraded, especially during restoration.

In 1973, when Morgan stretched the Plus 8's track by a couple of inches, Salisbury produced a much-needed tall axle ratio, up from 3.58 to 3.31:1 (the lower final drive could still be specified if you wanted sprint ratios). Naturally, when the SD1's all-synchro five-speed gearbox became available in 1977, Morgan adopted it after making just over 700 cars with Rover four-speeders. It came with an uprated (155bhp) SD1 engine (which no longer needed a dented air filter box), a new-look dash (vinyl-covered wood, crash-padded from '71), and 6-inch wide x 14-inch diameter wheels carrying 205/70 rubber, which meant

The leather-upholstered cockpit of the Plus 8 pictured on pages 76 and 77. The strip of warning lights just visible between the two main dials came from Jaguar's XJS.

widening the body, originally 56 inches (1,422mm) across, now 62in (1,575mm). It was at this time that an aluminium bumper replaced the old chromed one.

Enter the 'roller-skate' look, a significant landmark to Morgan fanciers. As fifth was effectively an overdrive – the mph/1,000rpm went up to around 26mph – and fourth much the same as the earlier gearbox's top, performance was not significantly affected by the ratio changes. Compared with the slow-shifting Moss 'box, which required a deliberate hand and accurate double-declutching, the shift quality of the two Rover gearboxes was quite reasonable, if a bit sticky. Motorway cruising and economy also benefited with the five-speeder – but steering effort increased as the width of the car and its tyres went up. Be warned: parking a Plus 8 calls for considerable muscle.

Changes to the original 3,528cc V8 engine initially centred on its cooling and exhaust systems. A more efficient water pump and a larger radiator, together with attendant plumbing improvements, resolved the problem of overheating in hot weather. Making the exhaust gases flow

The air cleaner of the earlier Rover V8 engines fitted to Plus 8s had to be slightly flattened in order to clear the Morgan's bonnet hinge, while the cars' right wing had to be cut away slightly to provide clearance for the alternator.

A 1983 Plus 8 owned by David Moss, carrying his private number-plate beneath the aluminium bumper. This car's smart appearance comes from metallic blue paintwork offset by light beige upholstery.

A Morgan for all occasions. The rollcage of this 1988/89 Plus 8 reveals the owner's serious intent, a tonneau cover and wind deflectors attached to the screen are aimed at maximum comfort, and there's a luggage grid on the tail. The Rover engine is supercharged by Brands Hatch Morgan. See page 86.

more freely also released a little more power. The compulsory introduction of an extra silencer (muffler) to meet noise regulations made for a rather flatulent exhaust note which rival TVR has always somehow avoided.

There was no publicity fanfare from Morgan when the Plus 8's SU carburettors (later Strombergs) gave way to fuel injection in a natural process of evolution dictated by emissions regulations, not to say Rover's quest for Vitesse potency. In view of the big hike in performance, why not the Plus 8 Super Sports, *Motor* magazine asked Peter Morgan in 1984? "Because the car has not yet *achieved* anything," came the reply. Engineering director Maurice Owen told *Motor* the main problem was "getting all the bits under the bonnet." These bits included the engine's management 'black box' bolted to the car's ash frame. Ancient and modern, in perfect harmony, as it were – but an end to simple DIY tinkering, too.

Before the adoption of the SD1 Vitesse engine, the Plus 8's power output, strangled by emissions regulations and the loss of five-star petrol, had been steadily falling. It started at 160bhp net, then dropped to 151 under the new DIN standard, and eventually fell to a miserable 143bhp (DIN) in '73, when the slowest Plus 8s were made. With

Another rollcage and a trio of badges point to a busy life for this 1974 four-speed Plus 8. Note triple wipers.

A 1983 Plus 8 with the tailpipes of the twin exhaust system just visible outboard of the reversing lights.The square lenses flanking the spare wheel are fog guards.

the SD1's launch in 1976 it increased to 155bhp. The big power hike, to 192bhp, came in 1983 with Lucas/Bosch electronic fuel injection (the extra 2bhp over the Rover Vitesse was down to Morgan's own exhaust manifold). Torque was also up, to 220lb ft from its 200lb ft nadir. Because of a non-standard close-ratio gearbox with high intermediates, *Motor* magazine was unable to match Morgan's standing-start performance claims. *Autocar* did. "It is not often we test a two-wheel drive car which reaches 30mph in two seconds...the benchmark 60mph comes up in 5.6 seconds."

To the best of my knowledge, no production Plus 8 has bettered that time. Despite a modest top speed of around 125mph, Morgan's mightiest was now in the supercar league – at a sub-supercar price of £13,000 in '84, when the Lotus S3 Esprit cost £15,985 and Porsche's 911 cabriolet

A 1985 fuel-injected 3.5-litre Plus 8 with the wide body and an interesting line in number-plates beneath the badge bar.

£25,556. Even TVR's like-powered Tasmin 350i cost £2,400 more.

The injected Plus 8 was also the first Morgan to get all-new rack-and-pinion steering, designed to minimize, if not eliminate, the notorious bump-steer unruliness of the superseded Cam Gears worm-and-nut affair. Unable to find a suitable proprietary rack and pinion with centre takeoff, Morgan commissioned Jack Knight to design and build one specially. *Motor* magazine had mixed feelings about the new setup, which increased lock-to-lock turns from 2.4 to 3.4. "No longer do your biceps need to be flexed as the Plus 8 is parked; it really is quite light. At speed, the steering is a great deal more precise as the inch or two of free play at the wheel rim has disappeared. On the debit side, it lacks feel...and at speed is lighter than desirable." All Plus 8s have collapsible steering columns, by the way.

Autocar described the injected Plus 8's ride as "diabolical", but conceded that it was "a joy to drive...a car

A unique two-seater drophead coupe version of a Plus 8 built up from a chassis, with the approval of the Morgan factory, by Bob Harper.

A burr walnut dashboard incorporating a lidded glove locker became an optional extra from 1989.

Morgan Plus 8s have continued to grow in width as well as stature. This prototype, built in 1991, has 2-inch wider wings to cover 16-inch chromed wire wheels.

which really separates the men from the boys." Gas-filled rear dampers were still 'experimental' at this juncture – though cars sold in the US had had them for years and Maurice Owen was in no doubt about their superiority. He told *Motor*: "You used to hit the first bump in a Morgan, miss the second one, and hit the third. Now you touch the road in between..." It was not until 1991, though, that the old lever-arm dampers were displaced by Gabriel telescopics, anchored at the top to a tubular hoop, which helped stiffen the chassis.

When Land Rover entered the US market with the Range Rover, Morgan once again had a petrol engine that satisfied US emissions regulations. It has been soft-selling cat-cleaned cars there since – doing away with the need for LPG-powered ones. In Europe, Germany was the first market to demand exhaust-cleansing catalytic converters. Cats are now obligatory for Moggies across the range, of course.

To compensate for the loss in power caused by tight emissions control, displacement of the Rover V8 was increased from 3.5 to 3.9 litres for the start of the Nineties, giving rather different delivery characteristics, but much the same overall level of performance. Power was little affected at 190bhp, but torque was increased by 7 per cent to a root-pulling 235lb ft at 2,600rpm on cars with catalysers – optional at first (when there were few takers), later standard. Afficionados assert that these rare non-cat 3.9s are the Plus 8s to have. Aurally, they're more aggressive with their twin-pipe exhausts, but the more subdued cat-car (with single-pipe exhaust) is actually just as powerful, and even torquier. The trouble is, it doesn't *sound* so exciting.

Top speed of the 3.9 still hovered around 120mph, which is fast enough in an open car, and the yardstick dash remained much the same at around 6 seconds. Pickup in fourth and top, geared to give an intergalactic 27.6mph per 1,000rpm, is where the Plus 8 showed the opposition a clean rump. As *Autocar* observed in its May '91 test, the 3.9 had the contemporary Porsche 911 Turbo well beaten up to about 80mph. Not many cars can live with a third-gear 50-70mph time of 3.1 seconds (it takes 4.3 seconds in fourth and a mere 6.3 seconds in top).

Said *Autocar*: "Plant your right foot hard down on the roller-type throttle pedal until it meets the floorboards – yes, it really is wood – and no matter what gear...the Morgan...launches itself at the horizon with an explosive surge..." *Autocar* observed that for "in-gear grunt" the Plus 8 had Porsche's Turbo beaten. Its testers also liked the noise (a muted bellow) and fuel consumption (over 20mpg) but, as always, they had reservations about the chassis. "The steering weights up challengingly as you fight against the understeer. Boot the tail out of line on the exit with a whiff of throttle – it seems quite natural with the Morgan – and you have to wrestle with the low-geared steering...this isn't a car for wimps. Total involvement is the message, sheer exuberance the key to its charm."

Civility came in for some stick, too: "Creature comforts are nil...deafening wind roar makes a radio pointless...although the comical triple wipers clear the screen reasonably well, the rest of the windows mist up badly." *Autocar* reserved its strongest criticism for the ride:

Apart from its dimensions, a Plus 8 can be recognized by the four sets of louvres cut into the top and sides of the bonnet. This 1991 car has a second rubbing strip on the mudguards beneath the doors, a popular option, especially on the wider-winged models.

This is the 1994 prototype of the long-door Plus 8, mounted on centre-lock alloy wheels shod with Pirelli P600 tyres. The car was sprayed in yellow and fitted with black upholstery and hood.

The flashing indicators have been outboard of the rear lights since 1989. In line with the significant improvement in manufacturing quality evident in recent years, a lot of attention has been paid to the hood.

"Every road joint is faithfully imprinted on your backside...damping control is poor, particularly at the rear, suspension travel inadequate and the rear axle hops and leaps..."

Had the Plus 8 retained its original sylph-like proportions, the 3.9 would be faster still. Over the years, though, Morgan's flagship car has grown considerably in girth and weight. While the wheelbase has remained constant at 92 inches (2,337mm), length has increased from 153in (3,886mm) to 159in (4,039mm), width from 58in (1,473mm) to 67in (1,702mm) with wire wheels. This made the cockpit less cramped for the corpulent, but it increased the drag of a car that, at best, has the aerodynamic efficiency of a railway station.

Increases in weight are no less significant, but harder accurately to quantify. Fuelled for 50 miles, *Motor*'s '68 test car, MMC 11, tipped the scales at 17.2cwt (1,926lb, 874kg), according to MIRA's weighbridge – used at the time by testers from *Autocar* and *Motor*. Nearly three decades later, Morgan's 1997 brochure quotes the Plus 8's

The dashboard of a 1989 Plus 8 with central rocker switches, speedo and rev-counter paired behind the wheel, and optional burr walnut. The steering wheel is not of standard design.

A tight fit beneath the bonnet for a Rover SD1 carburettor-fed 3.5-litre V8, this one supercharged by Rick Bourne of Brands Hatch Morgan. The 'blower', atop the engine, gives sensational acceleration.

The later Rover 3.9-litre V8 engine, without catalyst, installed in a 1992 Plus 8. Note the subtle cutaway of the bonnet hinge to provide clearance for the plenum chamber.

weight as 18.5cwt (2,072lb, 940kg) – though in real-life terms, it's probably more. While doing research for this book, significant discrepancies in quoted weights came to light, most independently weighed Morgans being heavier than the catalogue suggested – perhaps because of optional equipment. Assuming comparable all-steel bodies, the difference in weight between the original Plus 8 and a typical '97 one is probably greater than 1.3cwt.

As on the 4/4 and Plus 4, Girling brakes were displaced by servo-assisted Lockheed ones in 1993. The following year, the Plus 8 was also given Rover's new R380 gearbox (as was the Plus 4). Designed for the new Range Rover, this upgraded five-speeder was stronger, quieter and easier to use. First and reverse, adjacent on the old 'box, were now widely separated, minimizing the risk of wrong-slotting. The quality of the shift was also improved.

Rover's V8 was always a tight squeeze under the Plus 8's bonnet. In 1994, development engineer Bill Beck faced a new twist to the pint-in-a-quart-pot problem when Rover started delivering engines fitted with air-conditioning and power-steering pumps, both surplus to Morgan's requirements. Beck devised a new system of drive belts that catered for a water pump now rotating in the opposite direction. As a result of this work, access to the relocated alternator was improved. "You can now change it in a few minutes," said Beck. "It used to take an hour I should think."

To satisfy ever more stringent noise regulations, the Plus 8 gained a dual stainless steel exhaust system early in 1997, as did junior siblings. I put it to Bill Beck that, for a sportscar, the Plus 8's exhaust was already too muted, especially when compared to that of a burbling TVR powered by the same engine. "You should hear it now. It's really quiet," he said. Shortly after our conversation (in April '97), Beck was off to

Spain's official testing station near Barcelona to get the 4/4 and Plus 4 noise-certificated. Why Spain? Because noise measurement is affected by temperature, it seems. "Ten degrees is worth one decibel," said Beck. Other upgrades, shared with other models, include new rear axles (BTR for GKN), dust-free painting, and those cockpit revisions needed to install airbags (see Chapter 4).

The running changes made to the Plus 8 are too numerous to list in detail, but here are some of the more significant ones (chassis numbers are not indicated because, as outlined previously, they can be misleading):

October 1968
Morgan Plus 8 launched at Earls Court motor show, after supply of V8 engines had been delayed by the Leyland/Triumph merger.
1969
Chassis widened by 2 inches (but not the 49in, 1,245mm track); dual-circuit braking introduced; aluminium bodywork listed as an option, saving 100lb (45kg) in weight.
1971
Larger rear lights mounted in chromed-plastic tubes introduced on all models to comply with law. Floor reverts from metal to traditional wooden one.
1972
Four-speed, all-synchro Rover 2000-based gearbox, as used in 3500S, replaces old Moss unit. As box is now *in situ* with engine, a remote shift linkage is introduced. Enlarged radiator improves cooling. Stronger stub axles and steering arms fitted following breakages. By now, original single exhaust pipe has given way to two (exhaust system was to change several times after this).
1973
Front wings widened; fug-stirrer heater replaced by more efficient unit drawing fresh air; final drive ratio increased from 3.58 to 3.32:1; power reduced to 143bhp; track widened to 51/52in (1,295/1,321mm) front and rear. Body also widened.
1975
Sports Lightweight introduced. Extended wings accommodate wider wheels.

Morgans have never offered a Plus 6 model, but this one exists! It has a BMW 2.5-litre straight-6 engine coupled to a Getrag gearbox, an interesting transplant carried out by Colin Musgrove.

1977
Five-speed Rover SD1 gearbox introduced with 155bhp SD1 V8. Engine moved back in widened chassis – track now 53/54in (1,346/1,372mm), improving weight distribution; radiator enlarged; original rough-cast 5.5x15 alloy wheels replaced by 6.0x14 Milrace alloys carrying 205/70VR14 Dunlop or optional 195/70VR14 Michelin XWX tyres. New facia layout with speedo and rev-counter ahead of driver and supplementaries to centre (walnut veneer was not introduced until '89). Aluminium bumpers displace steel ones.
1978
Uprated servo-assisted brakes introduced.
1981
New purpose-built alloy wheels carrying Morgan logo introduced; back to 15 inches with 6in rims carrying 205/60 rubber. Brake servo dropped.

1982
Stromberg carburation for reduced emissions.
1983
Jack Knight rack-and-pinion steering announced as £1,100 option, later becoming standard on Plus 8.
1984
Injected Vitesse engine introduced, first as option, later standard. Power increases to 192bhp, torque to 220lb ft.
1986
Cuprinol protection for ash frame introduced as option, later standard. Other major measures which follow to combat corrosion include powder coating, galvanization and non-rust nuts, bolts and fastenings.
1989
First car with catalytic converter delivered to Germany.
1990
190bhp 3.9-litre Rover V8 introduced. Power unchanged, but torque increased. Available at first without catalytic exhaust. Later cars all with 'cats', giving even more torque.
March 1991
New Gabriel telescopic rear dampers attached to bracing tube over rear axle.
March 1993
Indian-made wire wheels offered as option for first time on Plus 8.
August 1993
New Lockheed-made braking system with servo and four-piston front calipers.
1994
Uprated Rover RT80 gearbox fitted giving improved shift quality. Belt drive for ancillaries revised, alternator tucked in and made more accessible.
1996
GKN rear axle displaced by cheaper Australian-made BTR one. Warranty claims drop.
1997
Introduction of long-door, deep-dash cockpit with more rearward seat movement to facilitate fitment of (optional) airbags. Exports to America, curtailed the previous year for safety and emissions control reasons, resumed. New low-noise stainless-steel exhaust system introduced.

In due course, the 3.9-litre engine (from the Discovery and Range Rover) will be supplemented, perhaps displaced altogether, by the high-torque, 220bhp 4.6-litre engine of the flagship Range Rover, bringing with it even greater performance. "We'll get it eventually," promised Peter Morgan late in 1996. He also asserted that Rover V8 supplies were secure for several more years. "They will definitely last into the new millenium, to 2002." By early 1997 a prototype 4.6-litre Plus 8 was already up and running, but the four-valve heads that Peter Morgan would like to have seen won't happen, but then he concedes the biggest advantage of Rover's two-valve V8 is its low cost.

The eventual demise of this evergreen engine is inevitable, of course. Despite continuous development, it is old, low-tech and, in the eyes of BMW, Rover's owners, surplus to requirements. You do not need to be clairvoyant to see that, in the long term, a 30-year-old GM cast-off has no place in the future of a progressive German manufacturer – especially when it has modern 3-litre and 4-litre V8s of its own. Not that these engines are likely powerplants for future Plus 8s. Morgan will be looking closer to home for cheaper alternatives.

CHAPTER 8

Morgans in competition

Flying the flag in trials, rallies and races

Morgan and motorsport go together like nuts and bolts. Right from the start, HFS Morgan promoted his three-wheelers by proving them in competition. Low prices, stiff suspension, agile handling and a good power-to-weight ratio attracted keen, sporting drivers as well as impecunious ones. For many trike owners, it was a small step from transport to trials. Beckoning beyond came hillclimbs, rallies, racing and record-breaking. In the thick of it were the Morgans themselves. Three generations of the family have been keen competitors in one field of motorsport or another.

When HFS first showed his single-seater three-wheelers at the 1910 motorcycle show, they'd done nothing of note to merit attention. Buyers in those days were swayed by prizes, not promises. A year later, Morgan could back his three-wheelers' performance and stamina with gold medals, won the hard way in reliability trials held over punishing terrain against the clock. As Bill Boddy observed in a 1965 issue of *Motor Sport*: "One of the secrets of the Morgan affair was the personal interest HFS took in competition driving. He was frequently to be found in trials, from 24-hour affairs to strenuous six-day marathons, accompanied by his wife Ruth." This tradition was continued by HFS' son Peter Morgan, who was a regular trials and rally competitor. Grandson Charles has maintained the family tradition in circuit racing.

To list all the three-wheelers' competition successes is beyond the scope of this book, but here's a summary of some of the more memorable ones:

1911
HFS Morgan establishes ruggedness and agility of Malvern-built three-wheeler by gaining gold medal in MCC London-to-Exeter reliability trial.

1912
HFS Morgan breaks one-hour cyclecar record, covering 59 miles and 1,120 yards in a JAP-powered single-seater Morgan.

1913
Gordon McMinnies, of *The Cyclecar*, wins French Cyclecar Grand Prix at Amiens in long-wheelbase JAP-powered Morgan. He averages 42mph over 164 miles. Production replicas, dubbed Grand Prix models, go on sale at £115.

1919
HFS Morgan wins 'gold' in ACU's Six Days trial in 6.8cwt 980cc Morgan trike. Built-to-order production Aero models soon follow.

1920
More 'golds' in Scottish International and Six Days trials for Morgan trikes. "The private owner can win on a Morgan," pronounced the advertisements. Note the word *on*, not *in*.

1922
EB Ware upholds Morgan's track record against motorcycle combinations, and Hawkes' 'Flying Spider' wins Brooklands race at 73.6mph. The Rev George Morgan claims that in previous 12 months, half-a-dozen Morgans between them won 24 'golds' and 12 'silvers' in major trials events.

From the start of production, four-wheeled Morgans were to be seen competing in trials. When things became tough there was usually plenty of manpower to augment the car's horsepower.

1923
EB Ware drives 715lb Morgan trike fitted with front-wheel brakes to victory in JCC (Junior Car Club) General Efficiency trial, averaging 56mpg and 55.7mph at Brooklands. Douglas Hawkes sets class speed record of 92.17mph in Anzani-engined Morgan.
1924
EB Ware again wins JCC General Efficiency trial, with best mpg, quickest acceleration, highest top speed and shortest braking distance. Harold Beart pushes class record to 97.32mph with ohv Blackburne engine.
1925
Morgans win four 'golds' and three 'silvers' on Lands End trial, another six 'golds' and two 'silvers' in London-to-Edinburgh trial. Harold Beart exceeds 100mph, doing over 104mph for flying-kilometre in special streamliner.
1926
Three-car team of water-cooled Aero JAPs win class team award in strenuous international 'Six Days'.

HFS Morgan, in the original drophead coupe, playing to the gallery as he fords a stream on the Crackington section of the 1938 Land's End trial.

A works 4/4, still fitted with standard two-seater bodywork, tackling a hill during the 1938 Land's End trial with HFS Morgan at the wheel. Subsequently, this car would be rebodied as a TT Replica.

The Geoffrey White/Dick Anthony Morgan rounding Arnage on its way to 15th place overall in the 1939 24-hours race at Le Mans.

1929
Ten Morgans start Edinburgh trial, nine win gold medals, including four-up Family model. Gwenda Stewart breaks several speed records at Montlhery in Hawkes-prepared, air-cooled 998cc Super Sports JAP. She pushes hour record to 101mph.
1930
Gwenda Stewart does over 113mph on her way to various class speed records. Morgans collect 11 'golds' in Lands End trial.

Summing up the prewar years, Bill Boddy wrote in *Motor Sport*: "For a simple vehicle, the keynote of which was utility, the Morgan three-wheeler gained a quite fantastic number of successes...its sponsors, of whom the Rev Morgan died in 1937 and HFS, at the age of 77, in 1959, purposely introduced the sport GP, Aero and Super Sports models and encouraged competition activities..."

Nothing much was to change when the four-wheeler was

introduced in 1936. The first 4/4s were immediately pitched into combat, especially in reliability trials – road sections interspersed with tricky tests against nature and the clock – that had given the tricars such a good name. In the quad ranks, though, the competition was stiffer. Even so, the 4/4 was soon accumulating awards in the Exeter, Lands End and Edinburgh trials.

Rallies in the Thirties were not the all-out, special-stage speed contests they were to become. Nevertheless, for Morgan's works manager (later managing director) George Goodall to win his class in the RAC rally in three consecutive years – 1937, 1938 and 1939 – was a fine achievement. Goodall's son, who regularly accompanied his father, won the Scottish rally in '39.

The 4/4 is believed to have scored its first race win in the 1937 Ulster Trophy, helped by a high rate of attrition among competitors. The following year saw a Morgan, privately entered by novice Prudence Fawcett, racing at Le Mans for the first time. With co-driver Geoffrey Wright, Miss Fawcett finished second in class and 13th overall, averaging 57mph for 1,373 miles. A year later, Wright was back, with Dick Anthony and more power, averaging over 64mph to finish 15th, albeit hundreds of miles behind the winning Bugatti.

In the early Fifties, when rallying was still as much a test of stamina and navigation between the sprints, hillclimbs and driving tests at which the agile Morgans excelled, 4/4s and Plus 4s were very competitive. Three-car Morgan teams took the team prize on the RAC rally in '51 against massive Jaguar XK120 opposition led by Ian Appleyard and his famous NUB 1. The same team (Peter Morgan, George Goodall and Dr W Steel) won again in '52.

Throughout the Fifties, Morgans – particularly Plus 4 Morgans – did well in rallying at club and national level; Jimmy Ray won the London rally twice in the early Fifties and there was a class win, and third overall, for John Spare in the '56 RAC. Decades later, in the Nineties, Morgans were to rediscover their competitive streak in retrospective events like the Pirelli Marathon and Historic RAC rally, that merged nostalgia and speed with a fair bit of

The extra power of a Plus 4 made it a popular successor to the 4/4 in competitions. Here, the stiffly sprung chassis becomes airborne while competing in the 1983 Coronation rally.

Lyndon Sims and Barry Hercock, class-winners on the 1960 Tulip rally with this Plus 4 drophead coupe, were doubtless pleased to have the extra protection of a hardtop.

socializing.

By the mid-Fifties, the Plus 4 was beginning to make its mark on both sides of the Atlantic. There were successes in the Sebring 12 Hours in '55 and '57, as well as at Watkins Glen, Fairchild and Lime Rock. The rorty Morgans were competitive against production class rivals – Bristols, ACs, MGs, TRs and so on – if not outright track winners. In rallying, hardy types like Pauline Mayman and Valerie Domleo (regular *coupe des dames* winners), not to mention Lyndon Sims and Brian Harper, flew the Morgan flag. But as rallying developed into rough-road racing as we know it today, so Morgans, ill-suited to this form of competiton with their cramped cockpits, poor ground clearance and crude suspension, evaporated from the scene.

Although Morgans were racing competitively in UK club events throughout the Fifties, it was not until late in the decade that the Plus 4 excelled, largely through the efforts of Chris Lawrence, a latterday Freddie Dixon, who could tweak cars as well as he could drive them. Appropriately, it was in the series for the Freddie Dixon Trophy that

Morgan were back at Le Mans in 1962, when Chris Lawrence and Richard Shepherd-Barron won the 2-litre class and finished 13th overall with this Lawrence-tuned Plus 4 Super Sports. The neat hardtop has a sensibly large rear window.

Lawrence and his famous Lawrencetune Plus 4 trounced everything in sight. Lawrence's prowess on the track prompted Morgan to introduce the aluminium-bodied, wire-spoked Lawrence-tuned Plus 4 Super Sports in 1960 – a genuine case of competition improving the breed (see Chapter 5).

Lawrence carried the Morgan torch abroad in the early Sixties. He broke Porsche's class lap record at the Nurburgring, much to the amazement of the Germans (who have been Morgan freaks ever since), and Lawrence's team-mate Richard Shepherd-Barron finished a three-hour race third overall at the daunting Spa-Francorchamps circuit, ahead of several Porsches.

At Le Mans in 1961, in what *Motor* described as "an ugly scene", the Lawrence/Shepherd-Barron Plus 4 was turned away by the scrutineers for being too old-fashioned. "Take it to the museum," the Brits were told. Rising above the snub, a determined Lawrence presented the car again in '62, this time with Morgan's backing and the scrutineers' approval. The rest is history. Averaging almost 94mph, the

One of three Lawrence-tuned Plus 4 Super Sports taking part in the 1,000-kilometres race at the Nurburgring. This hardtop model was shared by Pip Arnold and Robin Carnegie, who brought it home second in class.

Chris Lawrence's TOK 258 led a busy life, and is seen here leading the MGs of Alan Foster and Andrew Hedges in the Peco Trophy race at Brands Hatch in 1962.

little TR-engined Plus 4, looking incongruous on a track dominated by Ferrari 250GTOs (and won by a Ferrari 330LM at over 115mph), ran like clockwork to win the 2-litre GT class, albeit at a speed well below that of the 1600 Porsche victorious in the division below. Shepherd-Barron later told *Motor*: "The Morgan was very relaxing to drive and almost ideal on a circuit with a billiard-table surface like Le Mans. Being closed, it was noisier but also more comfortable than an open car...if you have to drive something hard for 24 hours it would be difficult to improve on the Morgan..."

Running a special Lawrencetune camshaft that, along with other modifications, gave an estimated 160bhp, the Morgan – famous now for its TOK 258 number-plate – was nudging 130mph down the Mulsanne straight. However, it was consistency and reliability, rather than outright speed, that brought it home in 13th place overall – the same position as Prudence Fawcett's Climax-engined 4/4 at Le Mans in 1938. At international level, the success of Lawrence/Shepherd-Barron in the 24 Hours still ranks as one of Morgan's finest achievements.

At home, Lawrence was still going strong in '62, setting 2-litre class lap records at every track contested. Other successes included Adrian Dence winning the Freddie Dixon marque championship in his Plus 4, and second place in the Silverstone Six Hours for Morgan, beaten only by a Jaguar trio.

In '63, Lawrence led a three-car Morgan team which included Pip Arnold and Bill Blydenstein, best remembered for his very fast Borgward and Vauxhall racers, to a class victory at Spa, averaging nearly 102mph for the long-distance marque race. Soon after came class success for Morgan at the Nurburgring.

But fast though they were, competition Plus 4s were always hampered by aerodynamics akin to a house brick's. They accelerated hard, handled and cornered well on their ultra-stiff suspension – at its best on smooth race tracks – and could out-brake anything. What they lacked was top speed, as Chris Lawrence had discovered at Le Mans. A more streamlined body might add 10-20mph to the Plus 4's Mulsanne maximum.

Enter, in 1964 – after Lawrence had rejected a works Porsche drive – the aerodynamic BP-backed Morgan SLR (for Sprinzel Lawrence Racing). Of the four that were built, by race/rally driver John Sprinzel and Chris Lawrence, three were based on Morgan Plus 4 chassis, the other on a Trumph TR4. All looked like a cross between an E-type Jaguar and a Lotus Elite, with sleek, wind-cheating aluminium bodywork (made by Williams and Pritchard) supported by light subframes. Surprisingly, the change of clothes did not contravene the rules and was worth 3 or 4 seconds a lap (Lawrence's class lap record at Goodwood stands to this day). Special Morgan-backed TR4 engines with Lawrencetune crossflow heads were developed for the SLRs, but they were abandoned after suffering from plug fouling and lubrication problems.

Were the SLRs true Morgans? It's a controversial point (as protests against their eligibility in marque racing testified), but these special lightweight coupes, built with Morgan's backing, were certainly very successful, witness the performance of Lawrence – following his recovery from a serious road accident the previous year – in the '65 Double 500 (later the BOAC 500) at Brands Hatch. Delayed by a chassis breakage while leading the first race, he won the second against strong competition. Third place at Spa, in a race packed with Porsches, was another highlight. One spin-off from the SLR was the later adoption on production Morgans of stronger stub axles.

Up to the mid-Sixties, Morgan four-wheeler successes included the following headliners gleaned from the motoring press:

1939
George Goodall wins his class on RAC rally for third consecutive year. Goodall junior wins Scottish rally.
1950
Peter Morgan, driving a new Vanguard-engined Plus 4, wins honours on Exeter trial.
1951
Three-car Morgan Plus 4 team (Morgan, Goodall and Steel) win team prize in RAC rally, beating Jaguar XK120 trio.

1952
Morgan repeats team success in RAC rally. Jimmy Ray wins London rally (and again in '53).
1956
JT Spare wins class and comes third overall (behind an Aston Martin and a Jaguar) in RAC rally.
1958
Morgan team pipped at post in Silverstone Six-Hour relay.
1959
Chris Lawrence victorious in 21 out of 22 races in his fast Lawencetune Plus 4 to win Freddie Dixon Trophy for Modsports cars. Lawrence also leads three-car Morgan team to victory in Six-Hour relay at Silverstone. Brian Harper wins London rally – and Welsh rally in '60.
1961
Lawrence surprises continentals at Nurburgring by breaking Porsche's 2-litre class lap record by 7 seconds in new Plus 4. Shepherd-Barron finishes second overall in three-hour race at Spa. Lawrence/Shepherd-Barron's Le Mans entry rebuffed.

Even the Plus 4 Plus found its way into competitions. Here Peter Morgan is tackling one of the sections on the 1964 Land's End trial.

The SLR Morgans had a particularly busy season in 1966, which included this wet day at Silverstone for the Martini meeting.

1962
Lawrence and Shepherd-Barron win class at Le Mans in Lawrencetune Plus 4, registration number TOK 258. They cover 2,256 miles at an average speed of 94mph. Poor streamlining restricts top speed to 130mph, best lap times to 110mph. Reliability and consistency rather than speed are keys to success. Morgans also finish first, second and third (Lawrence, Blydenstein, Arnold) in class at Spa and Nurburgring. Adrian Dence wins Freddie Dixon Trophy, and team Morgan (which includes Brian Redman) finishes second in Silverstone Six-Hour relay.
1964
Chris Lawrence and John Sprinzel launch Morgan SLR, with streamlined body and traditional underpinnings. Car is immediately several seconds a lap faster than 'classic' TR-engined Plus 4 and excels in Double 500 at Brands Hatch, among many other races.

The launch of the Plus 8 in 1968 marked a new and exciting era in Morgan's competition history, though the four-cylinder cars continued to pick up important awards. Morgans have since been campaigned at every level from road-going production cars carrying no more than obligatory safety equipment, to full-blown race-tuned specials, bearing little or no resemblance to anything you can buy from Malvern Link.

One such car was John Macdonald's wide-wheeled, supercharged Plus 8 (RUP 10M), said to have 550bhp under its massively scooped bonnet. Macdonald was also involved with the wide-bodied, Traco-engined Plus 8 intended for (but not raced at) Le Mans.

Another was Rob Wells' spaceframe Plus 8, built by Wells' Libra Motive agency in North London for Modsports (modified sportscar) racing. This fearsome machine took to the tracks in 1980 with Morgan's blessing and assistance. It had a purpose-built tubular chassis of much greater torsionally rigidity than the normal ash-braced ladder-frame affair, and was suspended all round by coils and wishbones that provided the superior wheel control needed to keep huge gumball slicks flat to the road. The Weslake-headed

Morgans were invariably serious challengers for Modsports honours, and this Plus 8 is on its way to another victory at Brands Hatch in September 1972.

That famous number-plate again, this time on one of Morgan's early Plus 8 racers, a pale blue car that became a prolific winner.

engine, good for over 300bhp, was set well back, almost mid-mounted in the chassis, in the interests of traction and handling balance.

Clothing the beast was a one-piece glassfibre body hinged at the tail so it could be lifted up and propped for easy access to the business end. Wells' Modsports Plus 8 was no more a true Morgan than a NASCAR saloon is a Ford or Chevrolet, but no matter. It did nothing to diminish the hairy-chested image of the Plus 8, and Morgan basked in its reflected glory. Lightened still further, Wells' rocket won the '81 STP Modsports championship, wearing the registration number MMC 3. (MMC 1, you may recall, was the original Plus 8 prototype. The first 'production' car was MMC 11 – still owned by Morgan when this was written.)

Wells' 'silhouette' racer was not the last of its kind. Charles Morgan was the driving force behind Morgan's controversial Plus 8 GTR campaigned in the '96 BPR Global Endurance Cup series. "It's very expensive," said

Rob Wells rounding Silverstone's Woodcote Corner in 1979 at the wheel of a Plus 8, loaned to him by Bob Stuart after Wells had crashed MMC 11 in the previous race. The hardtop and bonnet have been borrowed from the damaged car.

Peter Morgan, who was not wholeheartedly behind the project. "It's being used by Charles with a view to another (new) model. We have a big difference of opinion here..."

The Plus 8 GTR was the most ambitious (and expensive) works-backed car ever raced by Morgan. Co-driven by Bill Wykeham, Morgan's purpose-built Plus 8 bore little resemblance to any production model, other than in style. However, even with its special aluminium chassis, race-bred suspension, ground-hugging bodywork and massive rear wing, it struggled all season to match the pace of wilder rivals from Lotus and Marcos, never mind Porsche, McLaren and Ferrari – the big-buck stars of the series.

Morgan's 'Global' racer was rooted in an earlier radical one-off built with the help of Andy Rouse Engineering (of touring car fame). Its chassis, constructed from bonded aluminium honeycomb, carried wishbone front suspension and Ford Sierra-based semi-trailing rear arms. The car was then topped out traditionally, complete with ash frame, and run experimentally on the road for some 10,000 miles.

It was from the chassis of this prototype that Morgan's GT2 racer evolved, at Harvey Bailey Engineering, for circuit racing in '94. The following year, journalist Tony Dron, who had to his credit a string of wins and lap records in Colin Musgrove's Plus 8 racer, track-tested Morgan's 5-litre, 385bhp GT2 for *Complete Car* magazine. "The lap times showed this prototype is in a different league from the traditional Plus 8," wrote Dron. "The independent rear suspension is more effective than a live axle...traction out of corners is great." Charles Morgan spoke to Dron of production replicas costing £50,000. "We'll only be happy to produce it when the technology to manufacture it is really buttoned-up," Morgan is reported to have said.

Although in 1996 the Plus 8 GTR was capable of lapping the Brands Hatch GP circuit in under 1min 34sec – and that's quick – at Global Endurance level the Morgan was off the pace. In the first '96 race, at Paul Ricard, the Plus 8 GTR qualified 50th out of 51. In the final round, it qualified last. The best result – at Brands Hatch in September – was 12th in class. Mike Cotton, reporting on the series for *Motoring News*, summarized Morgan's '96

Rob Wells in MMC 3 holding a narrow lead over Bruce Stapleton's Group 4 car into Copse Corner at the Bentley Drivers' Club meeting at Silverstone in 1980.

season thus: "Very gallant, popular team but not competitive. Nice flag-waving exercise for Morgan enthusiasts around Europe."

Undaunted, Charles Morgan and Chris Lawrence – recruited as team manager to renew an old associaion – persevered. With a team of mechanics, they were overseeing the 1997 race-car's preparation when I visited the factory late in '96. Screens protecting their activities in a corner of the repair shop from prying eyes, never mind from cameras, underlined the project's secrecy.

Was the Plus 8 GTR's chassis the precursor of things to come? According to Morgan's '97 brochure, improving the breed is the sole aim of the racing programme. "This car (the Plus 8 GTR) is testing many components... to determine whether they are suitable for production. These include the chassis, drivetrain, suspension and wheels ... with a renewed commitment to international racing, the Morgan Motor Company acquires valuable knowledge and experience...the success of future Morgans will continue to depend on robust engineering and development with

The lift-up body raised on Libra Motive Racing's MMC 3 to reveal its spaceframe chassis and far-back engine.

Rob Wells' car, carrying the number MMC 4, in its 1983 specification, with sliding-pillar front suspension, K & N air filters and an improved front airdam.

Rob Wells in action again at Silverstone in 1990, with the disconnected front anti-roll bar just visible below the Morgan's left front wing.

MMC 11 came out of 'retirement' at the Syon Park Heritage Museum to be used by Rob Wells at Pembrey in 1992. He is seen here kicking up the dust on the exit from the hairpin.

feedback from the track."

Although Peter Morgan believed that a competition Morgan should be bred from the production car, not the other way about, he conceded that a modern chassis was a probability. Whether this would replace or complement the traditional sliding-pillar one is open to conjecture. But if Morgan has in its sights superior underpinnings, where better to evolve them than on the track – even though Morgan and cutting-edge technology do not on the face of it seem likely bedfellows?

Meanwhile, back to earth at club level, the five-class, factory-sponsored Morgan Challenge – for standard and modified Morgans of all ages – continues to attract healthy grids of Plus 8s. Started in 1985, the Challenge seems to embrace Morgan's fun-car ethos rather better than the deadly serious Global series.

Here's a summary of some Morgan competition highlights since the early Seventies:

1974
Robin Gray wins Modsports (modified sportscars) series in a Plus 8 and is second in Thoroughbred series in TOK 258, the ex-Le Mans car – still going strong, incidentally.

1975
Chris Alford's Britten-prepared 4/4, running a Minister-tuned Formula Ford engine, wins price-classed BRSCC Prodsports (production sportscars) championship with 17 class wins from 17 starts. In same series, John Britten is second in £4,000 class after several outright wins in his Plus 8.

1978
Bill Wykeham wins class in Prodsports series, driving Plus 8. Also drives much-modified Group 4 Plus 8 in various international events. Charles Morgan wins another Prodsports series in MMC 11.

1979
Charles Morgan and Rob Wells, driving same Plus 8

One of the most impressive cars usually on display at the Morgan factory is this 1994 Plus 8 5-litre racer, campaigned by Charles Morgan.

(MMC 11) prepared by Libra Motive, win classes in two Prodsports series.

1980
Rob Wells launches his factory-backed, Libra Motive-built Modsports Plus 8 with 300bhp dry-sump, Weslake-headed engine. Car carries historic MMC 3 number-plate.

1981
Rob Wells wins Modsports championship with MMC 3 – now almost unbeatable. Patrick Keen wins British Thoroughbred Sports Car championship in TOC 258.

1982
Steve Cole wins Prodsports championship with works-backed Plus 8. MMC 11, driven by Wells, Alford and Paul, wins Snetterton 24 Hours, covering a record 970 laps.

1985
Mary Lindsay wins first Morgan national championship – later the Morgan Challenge – in her Plus 8.

1989
Rob Wells, driving roadgoing Plus 8 with 4/4 chassis (ROB 8R), is first recipient of Peter Collins Tray as winner of Morgan Challenge, now sponsored by factory.

1990
Rob Wells wins Morgan Challenge again in ROB 8R. Bill Wykeham finishes fifth overall, and first in class, in Carrera Panamericana. The following year, his TR3-engined Super Sports Plus 4 finishes fourth overall. Rick and Jane Bourne, of Brands Hatch Morgan, win their class in restored '59 Plus 4 on Pirelli Marathon.

1991
Bournes' ex-Meredith racing Plus 4 finishes fifth on Historic RAC rally. The following year, it's second in Mitsubishi Classic Marathon.

1992
Chas Windridge, driving Class C Fiat-engined Plus 4, repeats his '91 win in Morgan Challenge.

1994
Charles Morgan campaigns special race-bred Plus 8 in GT2 sports class internationally with view to competing at Le Mans.

1996
Charles Morgan and Bill Wykeham mount gallant flag-waving campaign for Morgan in BPR Global Endurance GT Cup against formidable odds. Despite special race-car chassis and suspension, their Plus 8 GTR finishes the season as it starts – at the back of the field.

CHAPTER 9

The making of a Morgan

Craftsmanship, patience and dedication

How Morgans are made must be one of the motor industry's most open secrets. When I was at the Malvern Link factory, taking pictures for this chapter, visitors – some of them prospective owners witnessing the birth of their car – were wandering around the place as though it were one of the town's tourist attractions. As a living museum, it is. A video about the joys of Morgan ownership was repeating itself every 10 minutes or so at the back of a 'showroom' housing an array of Morgans, some owned by the works, others by customers. The tiny JAP-engined 1913 Runabout housed there looked as incongruous alongside a Plus 8 as the computers did in the Georgian offices across the way.

Morgan's place in Pickersleigh Road is a humble single-storey edifice of drab brick (no lovely Cotswold stone here). It has a sky-lighted, zigzag roof straight out of a Lowry painting, and a sloping floor that workers have cursed for 70 years. It is of no architectural merit and nothing appears to have been done to alter its appearance since the early Twenties, inside or out. The only evidence of modernization on site during my first visit was an enlarged visitors' car park and the bare steel skeleton of what was to become Morgan's new paintshop. More of this later.

Reception (to which all visitors are urged to report) is no more than a lowly cream-panelled lobby. Jaded pictures and memorabilia, including sign-written publicity placards listing the three-wheeler Runabout's achievements, 1910-20, festoon the walls. There's a portrait of HFS, Morgan's founder, by the door that leads into the works, and a display cabinet of models in one corner. If the Reverend Prebendary HG Morgan, one-time vicar of Stoke Lacy, had strolled through wearing top hat and high collar, I would not have blinked an eye. The place where Morgans are made is time-warped into the past.

Mark Aston, assistant managing director, was my guide. He looked too young to have worked at Morgan for 20 years, but he has an encyclopaedic knowledge of the place and its procedures. It's difficult to get in the odd grunt, never mind a question, once Mark is in full enthusiastic flow about Morgan production. Where to start? It doesn't much matter as the whole place seems crazily topsy-turvy, despite recent procedural changes made to improve productivity.

Let's start with the raw material that makes Morgans unique: the ash. About 5 tons of it is used by Morgan annually, most sourced from Britain, though Belgium was once a major supplier. Why ash? Because it's a tough hardwood with a high strength-to-weight ratio. So high, in fact, it anchors the seat belts without steel reinforcement. What's more it is quite flexible and not prone to splinter – an important safety point. With good reason are axe handles made of ash. Wheel spokes and axles, too, in the old days. On one of my visits, I saw in the repair shop the remains of a car, just identifiable as a Morgan, that had been in a head-on collision with a hefty van. Surely it was not a survivable accident? Wrong. The driver suffered leg injuries, but was not fatally hurt.

You could say that the body frame of a Morgan is 100

Two stacks of bare chassis frames, the larger of which these days are shared by the Plus 4 and Plus 8, the 4/4's being 2 inches shorter and narrower. The light-coloured ones are galvanized, the black ones powder-coated.

Some of approximately 50 wood components which will eventually form part of a Morgan's body frame. Each part is cut and shaped in batches of about 50 to ensure an adequate stockpile.

years old before it's even assembled. The ash is not from coppiced saplings, but from mature trees, planted in the nineteenth century. The planks are kiln-dried before going into Morgan's open-sided timber store, where they could be left seasoning for a year or more.

There are over 100 separate pieces of wood, each cut and shaped individually by saw, plane or router, in the frame of a four-seater Plus 4. The two-seater 4/4s and Plus 8s have slightly simpler structures with fewer than 100 components. Each section is made in batches of around 50, so there's always a stockpile. The rear wheelarches – the single biggest wooden components – are glued ash laminates, bent into shape in massive wooden clamps that appear old enough to have been designed by Noah. The 'oven' they're cured in for several hours looks like a cupboard. Correction: it *is* a cupboard.

By woodworking standards, if not machined steel ones, the carpenters work to quite fine tolerances. Frame assembly is not a team job: one craftsman is responsible for the whole structure (every frame bears its maker's identity). It is glued, jointed and screwed together in jigs. Depending on the model, the task takes 10-12 hours. When the frame is finished, and released from its clamps, it is soaked in a tank of commercial-grade Cuprinol wood preservative, then hung out to dry. Prior to 1986, the frames were surface painted with black creosote, obliterating the beauty of the wood's natural graining (not that you can see it, more's the pity, when the trim's in place), and providing little in the way of protection. Regularly exposed to wet weather, some parts of the body frames of pre-1985 Morgans – notably the sills and rockers – were likely to succumb to terminal rot within a few years. These days, they're reckoned to last much longer.

In the past decade, Morgan has taken major steps to improve quality and durability, even though it meant increasing prices. Peter Morgan credits his son with the initiative. "I was very cost-conscious, just like my father, who made cheap cars really well. What he enjoyed most was building cars down to a price. He was very good at designing simple things that worked." Peter Morgan is clearly a chip off the old block. "I thought it was more important to keep prices down than to get too sophisticated. Charles pressed for build quality. It's a case of the younger generation getting it right."

Zinc-coated steel coachbolts now secure the ash frame to the steel chassis – separated from each other by a modern waterproof membrane. Other fixtures and fittings, including the hood frames, sundry brackets, valances and so on, are either powder coated (Morgan has its own facility for this work) or galvanized. So are most of the chassis frames, made for Morgan nearby by ABT at Ross-on-Wye (Rubery Owen and Rockwell used to be suppliers). Although hot-dip zinc coating – the most thorough method – adds several kilograms to the car's weight, most customers prefer to have it in the interests of durability. When I was in the assembly shop, only one chassis had the supplier's alternative black powder coating – a sort of plastic skin that's electrostatically deposited as fine granules, then baked on in an oven. Powder coating remains effective against rust so long as the coating isn't perforated.

The chassis is a simple affair made up of Z-section side rails and crossmembers: since the Rover T16 engine displaced the M16, Plus 4s and Plus 8s have shared the same chassis, which is 2 inches greater in width and length than that of the 4/4. No-one at Morgan pretends that the structure of either doesn't flex, even after the recent insertion of a rear tubular hoop to support the telescopic dampers now fitted to all models. Indeed, flexing is seen as an adjunct to the car's suspension. Before the running gear goes in, the chassis is drilled to accept all the attachments, and the wooden floorboards (treated marine ply, not ash) inserted. A softwood panel is used to support the fuel tank, made in-house.

Fitting the engine/gearbox to the trestle-supported chassis is done with the help of a muscle-powered hoist. It's here that a four-man, job-swapping team also fits the front sheet-steel valances (which support the wings), cockpit bulkhead, suspension, brakes, brake lines and steering – Gemmer recirculating-ball on the 4/4 and Plus 4, Jack

There is much hand work in Morgan's sheet metal shop, where panels are individually tailored to a car from flat sheet. Here a bonnet takes shape.

Knight rack-and-pinion on the Plus 8. This work in what's officially known as the chassis erecting shop is a long-winded task taking many man-hours. A batch of 10 cars will spend a week here before the wheels go on. MIRA-tested, Wheels India chromed 'wires', supplied by Motor Wheels Services, are standard these days on the 4/4 and Plus 4, optional on the Plus 8; wheels from India and, since early '96, specially made back axles from Australia. "We got a better price and superior quality from BTR on the other side of the world than anywhere in Birmingham," Mark Aston told me.

Although major assemblies like the Rover/Ford engines, gearboxes and back axles (formerly from GKN) are bought in and modified where necessary to fit, Morgan makes – or at least machines – many of its own components. These include the rear brake drums, front discs, front stub axles, engine mounts, shackle pins, hubs, steering dampers, front suspension, hood frames, bonnet catches and many other sundries besides. Most are fabricated in a crowded machine shop – the last one you come to on the site's down slope – bereft of computers and robots. It's here, too, that various sub-assemblies like the front suspension are built up, ready for installation in the chassis assembly shop.

After the running gear is in place, on goes the ash frame – another week's work for a batch of 10 cars – to give the rolling chassis a recognizable outline. Next, it's to the deafening metal-bashing shop, which must be the noisiest place in Malvern Link, never mind at Morgan's. Ear defenders are much in evidence, and you have to shout to be heard above the *ra-ta-tat* hammering of steel upon steel, augmented by the blaring of radios – or should that be wirelesses? Morgan's workforce likes its pop music.

Simply put, it's in the sheet-metal shop that Morgans are skinned, either in steel or aluminium sheet – or a mixture of both. With the notable exception of the high-curvature, wire-edged wings – these were formerly hand-made by Manchester-based Eva Brothers, but now supplied by Eyres Forgings – panels are individually tailored from a flat cut-to-shape sheet, using hammer, mallet or roller, specially for the car under assembly. Not the last car, or the next one,

Chassis are supported on trestles in the assembly shop to provide a comfortable working height for coupling up the mechanical units.

The ash body frame takes shape. Clamps are used while the component parts are glued and screwed into place.

Painted bodywork, ash frame and steel chassis are united. Now, with the car on trestles, it's time to attach various component parts.

Working with metal can be a noisy business, so ear defenders are in evidence at Pickersleigh Road. Here the bars of a radiator grille are being carefully aligned as assembly nears completion.

which may be slightly different, but *this* one.

Sir John Harvey-Jones, the BBC's troubleshooter, flinched when he saw this highly labour-intensive procedure. What he failed fully to appreciate, it seems, was that traditional hand crafting is germane to the Morgan magic, a sales attraction that machine production would destroy. It also explains why you cannot buy over-the-counter Morgan body panels (wings excepted). In the event of crash damage, new bespoke bodywork is specially tailored for the job. Morgan's manufacturing process is sufficiently flexible to cater for such contingencies, if not quickly to satisfy impatient customers. It all takes time.

Edges are neatly doubled-up by turning them over and there's much trial-and-error fitting, especially of the bonnet, the louvres of which are stamped in after shaping. Another 10-car week goes by here in the sheet-metal shop. All three models – 4/4, Plus 4 and Plus 8 – proceed through these assembly stages together, incidentally, each car being put together according to a unique build sheet.

It's true what they say about Morgans: no two are exactly

With the help of an experienced pair of eyes and a high-speed sewing machine, another optional mohair hood takes shape.

alike, especially when you take into account the personal whims of buyers, who can choose from a long list of optional equipment and extras. These include special paintwork, leather trim, upgraded headgear, spare wheel cover, badge bar, towing bracket, bonnet strap, walnut dash, organ-type throttle, a clock, radio, speakers, even such basic items as undersealing, rustproofing and outside door handles. Door handles? "It's part of the tradition to offer them as extras," explained Mark Aston. So buyers beware: although they're modestly priced individually, optional extras can add thousands of pounds to the price.

When my photos were taken, Morgan was still using the paintshop in the original factory complex. The new one under construction over the way was to house an environmentally-friendly water-based system from ICI, which was supposed to be operational, running in tandem with the old system for a while, by the summer of '96. It wasn't. Six months on, the new low-line grey paintshop, looking a bit incongruous next to the old brick factory, was still idle. The delay was caused by a filtration system that failed to meet Morgan's specification, necessitating lengthy remedial work.

The driving force behind this investment, which Mark Aston reckoned would cost Morgan £350,000, was the Environmental Protection Act. However, a major bonus is that the new paintshop would release valuable production space where it's desperately needed in the main plant opposite. By now, what was the old paintshop should have become a new body fitting and panelling area. Result: less time spent pushing cars around from one area to another, and therefore greater production efficiency – poor though it remains.

From a quality standpoint, there was nothing wrong with the previous two-pack acrylic system, installed in 1986. It gave a deep, high-gloss finish that was more flexible than the prone-to-crack cellulose used earlier. It also gave customers a theoretical choice of 30,000 hues, though most opted for one of the five standard shades – Connaught Green (the most popular), Corsa Red, Indigo Blue, Royal Ivory or black.

Cars progressed through the old paintshop on 'slave' wheels, bonnets removed, masked up to prevent unwanted overspray. Wings were unbolted, but not totally detached, to get good paint cover at the joins. The body was then thoroughly cleaned to rid it of dirt, grease and rust, and hand sanded. Flexible epoxy-resin filler smoothed away blemishes, dents, ripples and welding scars inflicted during construction. Most of the subsequent wet-and-dry rubbing down was done by hand. Four or five layers of cream-coloured matt primer were then applied before the body was key-smoothed for the last time and sprayed twice with topcoat in a hot-air baking booth.

After painting, the cars are reunited with their wings (bonnets go on later, so easy access to the engine bay is retained) and coloured beading inserted between the wings and body, acting both as cosmetic highlighting and corrosion protection. It's at this stage that cars get their wiring looms, lights and catalytic exhaust systems – fitted to all 4/4s since the injected CVH engine, and to Plus 4s since

Nearing completion, this wire-wheeled Morgan is almost ready to receive its triple wiper blades, bonnet, grille and front bumper.

In the depth of winter, the Morgan factory has a sombre face, but behind the brickwork a great deal of dedication goes into the manufacture of a unique British sportscar.

the introduction of the T16.

Morgan's factory is a male-dominated place, witness the pin-ups that festoon the workshop walls. You will find women, though, at the sewing machines in the trim shop – a Seventies acquisition and not part of the original factory – where the cars get their upholstery, seats, hoods, carpets, windscreens, sidescreens, bright metal, badges and so on. If it moves or works, it will be fixed, checked and double-checked here. Black PVC is the standard trim material, but many customers these days opt for Connolly leather, which can be supplied in most colours. At least four skins are needed for a two-seater. The basic black hood (and it's very basic) is a crude, unlined PVC affair. The expensive German mohair alternative looks much classier but costs a lot more.

Chaos appears to reign in the combined trim and final assembly shop where part-finished cars are rolled to and

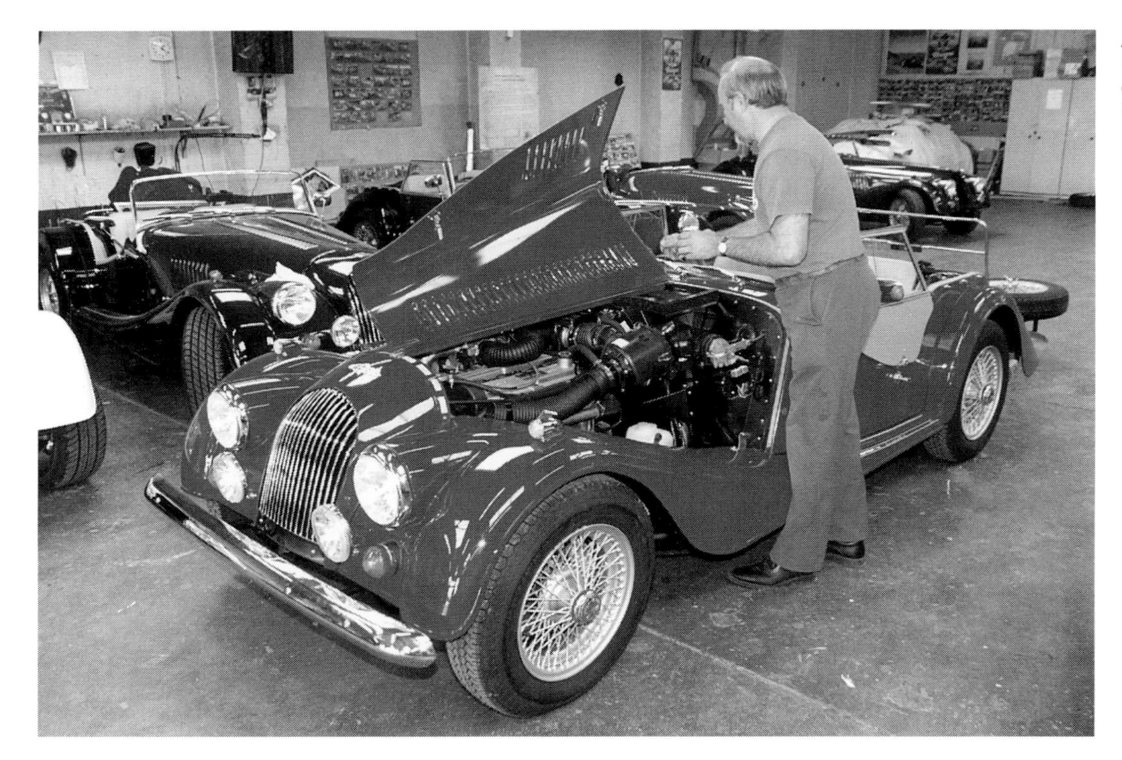

A final check-over and another Morgan is almost ready for its first owner. Only 10 or so cars leave the Malvern factory each working week.

fro, from station to station – and back again. Out of this muddle emerges two more-or-less complete Morgans every day of the working week, holidays excepted. Finally, away from the clean trim shop, the underwings are black-waxed. Surprisingly, full undersealing is still an option, as some customers don't want it. Every car is road-tested – Tony Monk has been doing the job for over 20 years – before it goes to the despatch shop for its bumpers and any last-minute touching-up or rectification.

When at last a new Morgan travels down Pickersleigh Road for the first time, either under its own power or on the back of a transporter, it is the end product of many hundreds of man-hours involving some remarkably archaic production techniques. It also carries with it a special type of heritage, one which is unique to the highly regarded products of Malvern Link.

CHAPTER 10

Buying and owning a Morgan

The choice, the inspection and fellowship

Morgans used to be scrapped, just like other cars, when they wore out – often prematurely. Not any more. Decrepitude is no longer cause for demise. Even the worst basket cases can be restored to *concours* standard – though most enthusiasts settle for a decent roadworthy condition. It follows that the world pool of Morgan cars, probably around 50,000 counting the three-wheelers, grows almost by the number of new ones made annually – say 470 at roughly 10 a week, allowing for holiday breaks. There's more good news: compared with Aston Martins, Ferraris, Porsches and other snooty sports exotica, Morgans are relatively cheap to buy and fix. The 4/4 is also reasonably economical.

Delivery times for new cars, real rather than apocryphal, stretched at their peak in the late Eighties to perhaps eight years, though they've since subsided to five or six. "They're part of the Morgan mystique," agent Richard Thorne told me. "High residuals depend on them. Good used cars will generally sell for the price that was paid for them new – perhaps even more."

Because each order for a new car takes so long to process, used Morgans hold their value exceptionally well. When demand exceeds supply, as it should according to HFS' creed, there's no cause for concern about depreciation – for most ordinary cars the biggest single cost. Morgans hold their own against inflation, even appreciate, so the cost of ownership can be surprisingly low. But beware. A ground-up rebuild by Morgan's repair shop – an adjunct to the factory – could cost almost as much as a new car, and take a long time to complete. By uniting DIY with professional assistance, though, restoration need not be cripplingly expensive, according to the experts.

You don't so much buy a Morgan as invest in one, as dealers (known as agents in Morgan-speak) will testify. So will speculators – very unpopular at Morgan – selling on their newly-delivered car at a premium price. In the late Eighties, when the market went mad and avarice prevailed, there were profits of £5-6,000 to be made on Plus 8s. Even though delivery times and values have since fallen (dropouts or deferrals help shorten the waiting list), a buoyant market is better than a slack one. Old hands at Morgan still remember the grim mid-Sixties, when demand was low and the factory had unsold cars lying around unwanted for years.

At the time of writing, Morgan listed 18 UK agents in its brochure, plus another four that provided service facilities. Just 23 agents covered the rest of the world, with two each in Germany and the US. All are listed in the appendix.

Prices of used cars vary enormously. When this book was published in 1997, a knackered 4/4 could be bought in the UK for under £5,000, if you were lucky. Runners ranged from under £7,000 for a roadworthy Ford-engined 4/4, Series I to V, to around £35,000 for a nearly-new, premium-priced Plus 8. Prepare for a long wait if you're particular about model and age, and remember that no two cars are identical. Many will reflect the bespoke whims of original owners who might perhaps have spotted something they fancied on a factory

The Morgan springing in close-up. The patented sliding-pillar front suspension, with coil main and rebound springs, has persisted throughout four-wheeler production. These days, modern piston-type dampers cushion the worst road shocks. At the rear the long laminated springs work in conjunction with angled dampers which are supported from a tubular hoop. Flex in the chassis frame makes its own small contribution to a Morgan's suspension.

visit (pilgrims are still welcome by Morgan at the Malvern plant to see their car being made, by the way). Others will have been tailored or upgraded by subsequent owners.

Just because Morgans are cherished worldwide, don't expect all old ones to be in pampered mint condition. Those made before 1986, which is something of a watershed in Morgan production history, were particularly vulnerable to the elements. In the bad old days, Morgans were not built to last. They were cheap sportsters, created for the enjoyment of first owners, not later generations. Build quality was poor, and little thought was given to durability in the production process.

Before the days of Cuprinol, powder coating, undercoating and galvanization, bodywork, chassis and frame were prone to premature – and quite possibly terminal – rot on high-mileage cars used year round, particularly on Britain's winter-salted roads. Garaging was not necessarily a preserving ploy, either, unless the atmosphere was dry. A dehumidifier is an old Morgan's best friend.

The finish of Morgans once left much to be desired, too. For instance, little attempt was made to titivate the engine bay, once daubed with a brush in black paint. And before '86, the crack-prone ICI cellulose paint was applied to

The driver of this badly mangled Morgan survived a high-speed head-on accident – a tribute to the integrity of a car designed long before the introduction of demanding crash-safety standards.

Morgan owners are served in the UK by a number of specialist dealers who offer extensive maintenance and repair facilities. This is a view inside the impressive Libra Motive premises.

completed cars over wing beading, so it soon became unsightly. When Morgan progressed to a more flexible ICI acrylic, it started painting the wings and panels separately, making sure invisible seams were coated. Modern waterproof piping sealed the gap 'twixt wings and body, greatly diminishing the risk of hidden decay.

Such practices may have been acceptable in the Sixties, even the Seventies, but by the Eighties the problem of quality, durability and finish was beginning to focus minds, particularly that of Charles Morgan, at Malvern. While the quality of mass-made mainstream cars had dramatically improved, Morgan's had marked time. After 1986, something approaching a revolution took place in a bid to improve the longevity. Unsatisfactory old practices were discarded, new ones put in their place.

Wood is a wonderful medium, with many great qualities, but there's a limit to what untreated BC (Before Cuprinol) ash can withstand when constantly bombarded with water, salt, muck and chemicals. Kept dry, a Morgan frame will last indefinitely. Left wet, it soon softens and crumbles, especially around the sills, doors and wheelarches. Wonky doors betray either frame rot or a chassis cracking. The ash frame can be repaired by replacing rotten sections, but extensive decay might mean a complete new structure, specially made either by Morgan (for patient souls who don't mind waiting) or one of several specialists. DIY restoration is quite feasible for the brave. One specialist, who reckoned the problem of rust was overstated, told me: "Morgans are not complicated cars – they're very repairable. You put them together like Meccano. A new body/frame can be bought for as little as £1,300."

Nothing more than a lick of paint once covered the bodywork, and the chassis was effectively unprotected. Any untreated frame more than 10 years old is almost bound to be rusty, quite possibly terminally. End of car? Certainly not. A replacement, galvanized or powder coated, can be supplied either by specialists or from the factory, though one repair shop told me chassis rot can usually be repaired by cutting out the bad sections and welding in replacements.

As with the original Lotus Elan, so with Morgans, a new lease of life is a new chassis, better protected than the original one was. Galvanization is not without its problems, however. For a start, it adds weight (though hardly significantly). Critics also contend that the zinc coat will eventually peel off under chassis flexing, and that rust radiates from the holes drilled into the frame members *after* zinc-coating. Waxoyl is a genuine life preserver on the inner faces.

Although aluminium bodywork was originally introduced to reduce weight and increase performance, it's usually specified these days to resist corrosion. Aluminium is not without snags, though: it dents more easily (thrown-up stones can damage the wings) and is prone to crack under flexing. And all Morgans flex. Where it's in contact with

Morgans ancient and modern in the company's display area provide a graphic reminder of how much, and yet how little, has changed over six decades of four-wheeler production.

The Morgan Sports Car Club (MSCC), which has some 4,000 members, provides owners with spares and technical support, as well as activities. This Plus 8 from Dubai attended Mog 94 in York.

steel, a chemically corrosive reaction can also set in.

Still, anything was better than untreated steel that rusted as quickly as the wood rotted – especially where the two were tacked together in damp union. The rule of thumb used to be simple: if it was made of unprotected steel, it would rust – sills, arches, inner wings, doors, battery box, bulkheads...they were all vulnerable to the dreaded red oxide on old Morgans, even some young ones if the original owner opted to skimp on protection. Panels and wings beyond repair can be replaced by aluminium ones, or even – and this might offend purists – glassfibre alternatives. Why not? The Plus 4 Plus had a plastic body, so there is a Morgan precedent. A wing replacement, incidentally, may well require a new bonnet (hood) even though the old one is undamaged. Remember, each and every bonnet is tailored to fit an individual car.

Basic information about every Morgan – chassis number, engine number, colour, date, despatch address and so on – can be checked against the factory's famous hand-written sales ledger. Over the years, though, Morgans tend to get modified, upgraded and refined in detail. Fittings and fixtures get changed – as they do during production, without notice or record to this day. The exact original specification of older cars is therefore difficult to verify. No matter. Authenticity, rather than the slavish pursuit of originality, will not offend purists. If it looks right, it probably is right. Originality is all very well, but who wants a Morgan rebuilt to its original poor standard? Why have an original component when a later authentic one does the job much better?

Another grey area that confuses owners and historians is Morgan's system of chassis numbering. Before the last War, chassis numbers were related to orders. So a cancellation meant a 'missing' number. When production resumed after the War, chassis numbers were allocated on despatch. This meant a car delayed during production, perhaps awaiting

Morgans galore. For owners, their car so often becomes central to their whole lifestyle, drawing them into membership of a huge family of fellow enthusiasts.

some special part, might eventually finish up with an out-of-sequence number – a system that led to some celebrated (or should that be notorious?) anomalies. Regard chassis numbers as a guide, no more.

Spares? The world's biggest source of Morgan bits is, appropriately, the Malvern factory. Parts manager Paul Trussler's main problem? "Space. We don't have enough of it." That's why Morgan's cut-off date for spares is 1950. "We have odd bits and pieces for prewar cars, but three-wheeler owners need to go to specialists." Dickensian though it looks, Morgan's profitable spares department is sympathetic to the needs of agents and owners, and therefore a key element in the company's heritage. What other manufacturer in the world can retail spares over the counter for cars half a century old? Paul Trussler told me that anything Morgan made in-house can be repeated, as the factory retains all its formers, templates and patterns – not to mention its expertise. "We can still do body frames for all the four-wheelers." The trouble with Morgan's made-to-order service is that, like the delivery of its cars, it's a tardy process that infuriates the impatient. "We are trying to improve things," said Paul Trussler. Meanwhile, outside specialists have cashed in on Morgan's sloth.

Many drivetrain parts for the early Coventry Climax and Standard engines, even the subsequent side-valve Ford ones, are difficult to obtain. For the later postwar cars, though, spares can generally be located from one source or another without too much trouble – though there are exceptions. Don't expect instant availability for Fiat 4/4 or 2-litre Plus 4, for instance. "Components are obtainable, but they can take some finding," said Paul Trussler. Rover's V8 engine has changed fundamentally in three decades and what's good for later engines is useless for early ones.

Although Morgans are basically simple cars served by proprietary engines (fitted, these days, with incongruous high-tech equipment necessary to meet ever-tightening noise and emissions laws), they will not tolerate neglect. However, most competent DIY mechanics would have little trouble with routine maintenance. The chassis – notably the steering and suspension – needs frequent attention with a grease gun to avert premature wear and unpleasant handling characteristics.

Although the sliding-pillar front suspension can be lubricated on the run with engine oil on postwar Morgans, aficionados say it's not the ideal lube. They also warn of messy overspill getting onto the brake discs. The job is best done with a gun-full of molybdenum disulphide-based grease. Regularly. Even then, the phosphor-bronze bushes in which the vertical pins slide wear quite rapidly, particularly on poor roads (the super-smooth roads of Germany help to extend suspension life). Typically, they need replacing every 15-20,000 miles or so to avoid steering flutter and disquieting clunking. Not surprisingly, kingpins are among the more popular lines of Morgan's spares department. The modern engines used by Morgan should be good for at least 100,000 miles given regular oil and filter changes. Major attention is unlikely to be unacceptably frequent on those Morgans – probably the majority – used mainly as second or third fun cars in the summer months.

Stiff steering usually betrays inadequate lubrication of the track-rod ends – again, a grease gun job. Not even purists are offended by the substitution of the old (and unsatisfactory) Burman steering box, prone to excessive play, by the later French Gemmer one. No Morgan has yet been built to my knowledge with power-assistance, so don't expect lightweight steering. Manoeuvring a wide-tyred Plus 8, which has rack-and-pinion steering, is hard work. Most Morgan rear axles can be replaced or rebuilt.

No Morgan owner – or prospective owner – should be stuck for advice or guidance. Apart from the agent oracles, there's also the Morgan Sports Car Club, established in 1951 as the Morgan 4/4 Club and democratically renamed in 1971 by postal vote. The MSCC now boasts some 4,000 members, over 700 of them living overseas. Through the club and its 30 regional centres in the UK, you can tap into a wealth of knowledge accumulated by kindred spirits over several decades.

Activities of the MSCC (president, PHG Morgan Esq) include the Morgan-sponsored Morgan Challenge racing

Morgans and motorsport are synonymous, and the connection has long been an international one. Here a group of competitors assemble in the paddock at Zandvoort, Holland, before heading on to the track.

championship (co-ordinated by Morgan works manager Mark Aston's wife Serena); hillclimb and sprint championships; an annual *concours* and dinner/dance; and a four-day international rally (when this was written, members were being polled about a plan to hold Mog98 – the 1998 grand gathering – in the Isle of Man, at £450 per couple; Mog97 was being held at Bath). At a local level – SexMog (Brighton), BogMog (Bristol), FellMog (Cumbria), LinMog (Lincoln), SmogMog (London) and many other OddMogs – club members can also enjoy treasure hunts, autotests and scenic runs in their Morgans. You will find MSCC members active elsewhere, too – in historic rallying, classic trials and production car trials, for instance.

Other benefits offered to members by the MSCC include a spares operation for older cars, favourable insurance premiums and RAC membership, a bookshop, and technical advice for all models. The Club's monthly 52-page magazine, called *Miscellany*, is essential reading for all Morgan owners and buffs. Technical experts, specialists, regional centres, affiliated clubs, events, reports, features, news, models, DIY, touring, regalia, for sale and wanted...you will find them all in the black-and-white A4-sized pages of *Miscellany*. You will find more Mogs for sale here than in any other publication, by the way: I counted 78 in the October '96 issue, to take just one at random. Runners ranged from £8,995 for a 68,000-mile 1982 4/4 (condition unstated but clearly not *concours*) to a steep £34,500 for an ally-bodied '96 Plus 8 3.9 with undersealing, roll bar, walnut dash and chrome wires. At the time, a standard Plus 8, bereft of options, was listed by Morgan at £29,328. The majority of cars fell in the £12-22,000 bracket.

In the same issue, an advertisement for Heart of England Morgans, well-known Moggie specialist, hinted at the wide range of parts available to owners by mail order. You need a valance tie-bar? It's yours for £25 from HoEM. Competition is keen in Morgan's small, tight-knit world, as the back-cover ad for Melvyn Rutter – another well-known Morgan sales, service, parts and restoration specialist – indicated. Rutter's stainless steel hood frames, £99 plus VAT, caught my eye. Rick Bourne's Brands Hatch Morgan agency was offering a supercharger conversion for the Plus 8, among many other things.

The factory is not the only source for a new Morgan chassis. GEE Ltd, for instance, was advertising own-make, powder coated chassis from £445. Stainless steel exhaust systems from Harpers cost from £125 to £695 plus VAT. In the same issue, from Phaeton Garage, south coast Morgan specialists, I spotted a 'remote upper kingpin greasing kit' to replace Morgan's messy oiler system. There were also two pages devoted to MSCC's spares service for older cars. If you want your Morgan to look nicer, handle better or go faster, there's a specialist out there who can oblige.

The MSCC is not the only club catering for Morgan owners. The Morgan Three Wheeler Club was founded in 1945, six years before the MSCC came into existence. Reflecting the amazing worldwide affection for Morgans, there are some 35 other dedicated clubs around the world. At the time of writing, there were no fewer than 15 in the US. Others can be found in Australia, New Zealand, Canada, South Africa, Japan and in every country of western Europe.

You're never alone with a Morgan.

APPENDIX A

Morgan milestones: who, what, when, where

1881: Harry Frederick Stanley Morgan – HFS for short – is born at Morton Jeffries, England.

1899: HFS joins the Great Western Railway as a trainee draughtsman.

1906: HFS opens garage at Malvern Link in Worcestershire, England.

1908: HFS starts construction, with help from Stephenson Peach, of Peugeot-powered trike, dubbed Runabout. Lightweight single-seater finished in following year.

1910: Morgan production starts. Tiller-steered Runabouts displayed at Olympia motorcycle show. Lots of interest, but few orders.

1911: HFS wins gold medal in Exeter trial. Two-seater with proper steering wheel on show at Olympia attracts so many orders HFS considers contracting out assembly.

1912: Morgan Motor Company formed, HFS marries and does 60 miles in an hour at Brooklands. Events unconnected.

1913: Gordon McMinnies wins French Cyclecar GP at Amiens in JAP-powered Morgan.

1914: Experimental four-wheeler made in response to Morgan's exclusion from certain events.

1915: Family four-seater, designed earlier, goes into production.

1919: New factory at Pickersleigh Road, Malvern Link, gives production potential of 2,500 cars a year.

1920: Quickly detachable rear wheel and equal-length chains introduced.

1925: Maurice Goodall joins Morgan as general manager, Harold Beart does over 100mph.

1926: Electric starting introduced to trikes.

1930: Mrs Gwenda Stewart does over 117mph on tree-lined road at Arpajon, near Montlhery.

1931: New chassis with three-speed (and reverse) gearbox and one drivechain introduced.

1933: 933cc four-cylinder side-valve engine from Ford Y-type introduced to new F-type trike in response to new four-cylinder BSA trike.

1935: Four-wheeler Morgan seen testing at Brooklands. New 4/4 announced late December.

1936: Improved production 4/4 appears at Olympia motor show. Fire destroys bodybuilding shop.

1937: HFS takes back seat as chairman, Maurice Goodall appointed managing director. Goodall wins class in RAC rally (and repeats success in '38 and '39). The Rev George Morgan dies on November 10, aged 86.

1938: £225 4/4 four-seater launched, followed by drophead coupe.

1939: New Standard ohv engine announced for 4/4. Only a handful made before War. Ford V8-engined Morgan, foreshadowing later Plus 8, abandoned. Ditto Arnott supercharger.

1945: Morgan Three Wheeler Club formed.

1946: Production of Standard-engined 4/4 resumes and Morgan looks to markets abroad.

1947: PHG Morgan, founder's son, joins family firm as development engineer and draughtsman.

1950: 2,088cc, 68bhp Standard Vanguard engine launches Plus 4. 4/4 dropped.

1951: Morgan 4/4 Club formed, later to become Morgan Sports Car Club.

1952: Last three-wheeler leaves factory in February, made previous year.

1954: Facelift, started previous year, sees finalized version of faired-in headlights and curved cowl over hidden radiator. Plus 4 gets 90bhp Triumph TR2 engine.

1955: Morgan 4/4 tourer re-introduced with side-valve Ford engine, price £713. First Morgan to get new sloping tail.

1956: TR2 engine replaced by 100bhp TR3 engine in Plus 4.

1958: PHG Morgan becomes joint managing director.

1959: Founder HFS Morgan dies in June, shortly before company's 50th anniversary.

1960: Following racing successes in late Fifties, Morgan introduces Plus 4 Super Sports with Lawrencetune TR engine.

1962: Lawrencetune-prepared Morgan Plus 4 wins class and finishes 13th overall at Le Mans. Series V 4/4 introduced with 1.5-litre ohv Ford engine.

1963: Plus 4 Plus with 'modern' glassfibre body supplements traditional Plus 4 at Earls Court show. Model sells poorly, only 26 made.

1968: Plus 4 replaced by Rover V8-engined Plus 8. Maurice Owen responsible for prototype and development, later becomes engineering director.

1972: Robin Gray drives Plus 8 to victory in Freddie Dixon Modsports championship.

1976: Chris Alford wins BRSCC Prodsports championship with unbeatable 4/4.

1978: Charles Morgan, founder's grandson, wins BRSCC Prodsports championship in MMC 11 (second prototype) with nine wins and three seconds from dozen starts.

1982: Steve Cole wins Prodsports championship and Rob Wells Modsports series with fearsome spaceframe lightweight Plus 8.

1984: Gathering of 1,200 Morgans at Eastnor Castle to celebrate company's 75th anniversary. Fuel injection adopted for Plus 8.

1985: Plus 4 re-introduced with 2-litre Fiat Twin-Cam engine (after 1.6-litre Fiat displaced by Ford CVH 1.6 in 4/4). Charles Morgan joins family firm full-time. Race series for Morgans starts.

1986: Start of quality revolution when Morgan addressed problems of corrosion and durability with galvanization, powder coating, Cuprinol dipping and so on.

1987: 2-litre Rover M16 twin-cam with 24 valves displaces Fiat engine in Plus 4.

1989: Plus 8 gets 3.9-litre engine.

1990: *Troubleshooter* Sir John Harvey-Jones tells Morgan they've got it all wrong. Morgan rejects his recommendations to invest, expand and modernize.

1991: Rover T16 engine replaces M16 in Plus 4. Performance is unaffected.

1993: Ford 1.8-litre Zeta twin-cam engine yielding 121bhp introduced for 4/4.

1997: New environmentally friendly paintshop in operation after delays caused by faulty air extraction.

APPENDIX B

Performance figures

Model	Engine (maker)	Approx. production dates (from-to)	Capacity (cc)	Power (bhp)	Power to weight (bhp per ton)	Gears (number)	0-60mph (sec)	Maximum (mph/kmh)	Source
4/4 S I	Coventry Climax	March '36 to Sept '39	1122	34	50	4	28.4	75/121	*The Autocar*
4/4 S I	Standard Special	March '46 to Feb '51	1267	40	56	4	26.0	77/124	Morgan
4/4 S II	Ford 100E	Oct '55 to Nov '60	1172	36	52	3	29.4	71/114	*The Autocar*
4/4 S II	100E Aquaplane	May '57 to Sept '60	1172	44	64	4	20.5	80/129	*Road & Track*
4/4 S III	Ford 105E	Oct '60 to Nov '61	997	39	56	4	25.8	78/126	*Car and Driver*
4/4 S IV	Ford 109E	Oct '61 to March '63	1340	54	83	4	18.6	80/129	*The Motor*
4/4 S V	Ford 116E	Feb '63 to March '68	1498	65	94	4	16.5	80/129	*Road & Track*
4/4 S V	Ford 116E GT	Feb '68 to May '71	1498	78	104	4	11.9	94/151	Morgan
4/4 1600	Ford Kent GT	May '71 to March '82	1597	96	126	4	10.0	103/166	*Motor*
4/4 1600	Fiat 1.6 TC	May '81 to Nov '85	1584	98	135	5	9.8	104/167	Estimated
4/4 1600	Ford 1.6 CVH	March '82 to Nov '91	1597	96	132	4 and 5	9.9	104/167	Estimated
4/4 1800	Ford 1.8 Zeta	March '93 to date	1796	121	165	5	8.9	118/190	Estimated
Plus 4	Vanguard	June '51 to June '58	2088	68	87	4	14.1	85/137	*The Motor*
Plus 4	TR2	May '54 to June '56	1991	90	109	4	13.3	96/155	*The Autocar*
Plus 4	TR3	March '56 to Dec '62	1991	100	121	4	9.7	100/161	*The Motor*
Plus 4	TR Lawrencetune	Feb '61 to May '68	2.0 & 2.2	115-125+	160	4	7.6	115/185	Morgan
Plus 4	TR4 & 4A	Oct '62 to Jan '69	2138	100-104	126	4	9.3	110/177	*Car and Driver*
Plus 4	Fiat 2.0 TC	April '85 to Jan '87	1995	122	146	5	8.8	112/180	Estimated
Plus 4	Rover M16/T16	Jan '89 to date	1994	138-134	162-157	5	7.7	110/177	Estimated
Plus 8	Rover V8	May '68 to Jan '76	3528	168	195	4	6.7	125/202	*Motor*
Plus 8	Rover V8	Oct '76 to Oct '83	3528	155	168	5	7.1	122/196	Estimated
Plus 8	Rover V8	June '90 to date	3946	190	206	5	6.1	122/196	*Autocar*

APPENDIX C 4/4 technical specifications

Model/years	SI, 1936-39	SI, 1946-50	SII, 1955-60	SIII, 1960-61	SIV, 1961-63	SV, 1963-68
Engine	4-cyl Coventry Climax	4-cyl Standard	4-cyl Ford 100E	4-cyl Ford 105E	4-cyl Ford 109E	4-cyl Ford 116E
Block/head	iron/iron	iron/iron	iron/iron (alloy comp)	iron/iron	iron/iron	iron/iron
Bearings	3	3	3	3	3	3
Bore/stroke, mm	63 x 90	63.5 x 100	63.5 x 92.5	80.96 x 48.41	80.96 x 65.07	80.96 x 72.74
Capacity	1122cc (some 1098cc)	1267cc	1172cc	997cc	1340cc	1498cc
Compression	6.8:1	7.0:1	7.0:1	8.9:1	8.5:1	8.3:1
Valvegear	oh inlet, side exhaust	ohv pushrod	sidevalve	ohv pushrod	ohv pushrod	ohv pushrod
Induction	Solex carb	Solex carb	Solex carb	Solex carb	Zenith carb	Zenith/Weber GT
Power/rpm	34bhp at 4500	40bhp at 4300	36bhp at 4400	39bhp at 5000	54bhp at 4900	60bhp/78bhp GT
Torque/rpm	Not known	61.6lb. ft at 2500	54lb.ft at 2500	52.9lb. ft at 2700	74lb. ft at 2500	81lb. ft/90lb ft GT
Transmission	Front engine, rear drive, central gearbox		Front engine, rear drive, gearbox attached to engine----------------------------------			
Gearbox	4-speed Meadows	4-speed Moss	3-speed Ford	4-speed Ford	4-speed Ford	4-speed Ford
Chassis	2-section side members with cross bracing. Ash frame, Cuprinol-treated from '86, supports steel or steel/aluminium					
Steering	Burman Douglas cam and peg -------►		Cam Gears cam and peg --			
Front suspensn	Sliding stub axle on braced upright pillar, coil springs and Newton hydraulic dampers, later Armstrong --------------------------					
Rear suspensn	Live axle, sprung and located by multi-leaf springs. Hartford friction dampers replaced by Armstrong piston dampers at SIII,					
Brakes	8in Girling drums, mechanically operated --►		9in drums, hydraulic operation -----►		11in discs/9in drums, hydraulic	
Wheels/tyres	Disc/5.00 x 16	Disc/4.50x17 & 5.00x16	Disc/5.00 x 16	Disc/5.20 x 16	Disc/5.60 x 15	Disc/155 x 15
Dimensions*						
Length, in/mm	140/3556	140/3556	144/3658	144/3658	150/3810	150/3810
Width, in/mm	54/1372	54/1372	52/1321	56/1422	56/1422	50/1422
Height, in/mm	52/1321	52/1321	49/1245	51/1295	50/1270	50/1270
W/base, in/mm	92/2337	92/2337	96/2438	96/2438	96/2438	96/2438
Track F/R, in/mm	45/45, 1143/1143	45/45, 1143/1143	45/45, 1143/1143	47/47, 1194/1194	47/47, 1194/1194	47/47, 1194/1194
Weight, lb/kg	1582/718	1590/721	1540/698	1540/698		1540/698
Price new, from	£194 5s (1937)	£456 15s 1d (1946)	£713 17s (1956)	£737 15s 10d (1961)	£774 3s 1d (1962)	£877 (1968)
Production	approx 900	539	386	58	114	639
Body styles	2 str, 4 str, DH Coupe	2 str, 4 str, DH Coupe	2 seater only	2 seater only	2 seater only	2 seater only

*All figures for 2-seater models

1600 Kent, 68-82	**1600 Fiat, 81-83**	**1600 CVH, 82-91**	**1600 EFI, 1991-93**	**1800 EFI, 1993-onwards**	
4-cyl Ford Kent	**4-cyl Fiat TC**	**4-cyl Ford CVH**	**4-cyl Ford EFI**	**4-cyl Ford Zeta**	**Engine**
iron/iron	iron/iron	iron/alloy	iron/alloy	iron/alloy	**Block/head**
5	5	5	5	5	**Bearings**
81.0 x 77.6	84.0 x 71.5	80 x 79.5	80 x 79.5	88.0 x 80.6	**Bore/stroke, mm**
1598cc	1584cc	1597cc	1597cc	1796cc	**Capacity**
9.0:1	9.0:1	9.0:1	10.0:1	10.0:1	**Compression**
ohv pushrod	2 ohc, 8 valves	sohc, 8 valves	sohc, 8 valves	2 ohc, 16 valves	**Valvegear**
Zenith/Weber GT	Weber or Solex	Weber carb	Electronic fuel injection	Multi-point fuel injection	**Induction**
70bhp (80bhp GT)	98bhp at 6000	96bhp at 6000	100bhp at 6000	121bhp at 6250	**Power/rpm**
97lb. ft (100lb. ft GT)	96lb. ft at 4000	98lb. ft at 4000	102lb. ft at 2800	120lb. ft at 4500	**Torque/rpm**
→	→	→	→	→	**Transmission**
4-speed Ford	5-speed Fiat	4 & 5-speed Ford	5-speed Ford	5-speed Ford	**Gearbox**
bodywork. Galvanization optional from '86, powder coating standard '86 on					**Chassis**
→	→	Gemmer recirculating ball from 1984 to date	→	→	**Steering**
→	→	→	→	→	**Front suspensn**
later gas-filled dampers					**Rear suspensn**
operation →	→	→	→	→	**Brakes**
Discs or wires/5.60 x 15, 165/15, 195/60	→	→	Wires/165TR15, 195/60VR15	→	**Wheels/tyres**
					Dimensions*
150/3810	150/3810	150/3810	153/3886	153/3886	**Length, in/mm**
56/1422	56/1422	57/1448	57/1448	57/1448	**Width, in/mm**
52/1321	52/1321	52/1321	52/1321	52/1321	**Height, in/mm**
96/2438	96/2438	96/2438	96/2438	96/2438	**W/base, in/mm**
47/47, 1194/1194	47/47, 1194/1194	47/47, 1194/1194	48/49, 1219/1245	48/49, 1219/1245	**Track F/R, in/mm**
1560/708	1600/726	1625/737	1700/771	1914/868	**Weight, lb/kg**
£1569 (1974)	£7413 (Aug '82)	£12,645 (May '90)	£15,483 (July '92)	£18,794 (Sept '96)	**Price new, from**
3513	96	2222	195	approx 400, early '97	**Production**
2 seater, 4 seater	2 seater, 4 seater	2 seater, 4 seater	2 seater, 4 seater	2 seater only	**Body styles**

APPENDIX D Plus 4 technical specifications

	1950-58	1954-56	1956-62	1962-69
Engine	**Standard Vanguard, 4-cyl**	**Triumph TR2, 4-cyl**	**Triumph TR3, 4-cyl**	**Triumph TR4/TR4A, 4-cyl**
Bore/stroke, mm	85 x 92	83 x 92	83 x 92	86 x 92
Capacity	2088cc	1991cc	1991cc	2138cc
Compression	6.7:1	8.5:1	8.5:1 (9.1 SSports)	9.0:1
Induction	Single Solex	Twin SU	Twin SU	Twin SU or Stromberg
Power/rpm	68bhp at 4300	90bhp at 4800	100bhp at 5000	100 at 4600 and 104 at 4700
Torque/rpm	112lb. ft at 2300	115lb. ft at 2600	117lb. ft at 3000	127 at 3350 and 132 at 3000
Block/head	iron/iron	iron/iron	iron/iron	iron/iron
Valvegear	ohv pushrod	ohv pushrod	ohv pushrod	ohv pushrod
Transmission	Front engine, rear-wheel drive via short shaft to central Moss gearbox, all cars up to TR4A			
Clutch	Borg & Beck s.d.p.	Borg & Beck s.d.p.	Borg & Beck s.d.p.	Borg & Beck s.d.p.
Gearbox	4-speed Moss	4-speed Moss	4-speed Moss	4-speed Moss
Chassis	Z-section side members braced by cross members, powder coated from '86, galvanization optional from '86			
Steering	Burman/Cam cam & peg. Later, Gemmer recirculating ball. Knight rack & pinion optional from '91			
Front suspension	Sliding stub axles on rear vertical pillars, coil springs with hydraulic dampers			
Rear suspension	Live axle sprung and located by semi-elliptic leaf springs, hydraulic dampers. Rear telescopics, mounted on bracing			
Brakes	Drums/drums. Optional front discs from '59, standard '60 →			
Wheels	16in discs, 15in discs or wires from '59 →			
Dimensions*				
Length, in/mm	140/3556	140/3556	140/3556	144/3658
Width, in/mm	56/1422	56/1422	56/1422	56/1422
Height, in/mm	46/1168	51/1295	51/1295	51/1295
W/base, in/mm	96/2438	96/2438	96/2438	96/2438
Front track, in/mm	47/1194	47/1194	47/1194	47/1194
Rear track, in/mm	47/1194	47/1194	47/1194	47/1194
Weight, lb/kg	about 1800/816	about 1850/839	about 1850/839	about 1850/839
Body styles	2 str, 4 str, DH Coupe	2 str, 4 str, DH Coupe	2 str, 4 str, DH Coupe	2 str, 4 str, DH Coupe
Production	799, 1950-58	344, 1954-56	1808, 1956-64	1582†, 1962-69

*All figures for 2-seater models   †including Plus 4 Plus

1985-88	1988-92	1992 to date	
Fiat TC, 4-cyl	**Rover M16, 4-cyl**	**Rover T16, 4-cyl**	**Engine**
84 x 90	84.5 x 89.0	84.5 x 89.0	**Bore/stroke, mm**
1995cc	1994cc	1994cc	**Capacity**
9.0:1	10.0:1	10.0:1	**Compression**
Fuel injection, Bosch	Fuel injection, Lucas	Fuel injection, Lucas	**Induction**
122bhp at 5300	138bhp at 6000	134bhp at 6000	**Power/rpm**
129lb. ft at 3500	131lb. ft at 4500	131lb. ft at 4500	**Torque/rpm**
iron/alloy	iron/alloy	iron/alloy	**Block/head**
2 ohc, 8 valves	2 ohc, 16 valves	2 ohc, 16 valves	**Valvegear**
Clutch & gearbox in unit with engine from 1985			**Transmission**
Fiat s.d.p.	Rover s.d.p.	Rover s.d.p.	**Clutch**
5-speed Fiat	5-speed Rover	5-speed Rover	**Gearbox**
------------►			**Chassis**
------------►			**Steering**
------------►			**Front suspension**
hoop from 1992 ------------►			**Rear suspension**
11in front discs, 9in rear drums ------------►			**Brakes**
Centre-lock 'wires' carrying 195/60 VR tyres ------------►			**Wheels**
			Dimensions*
153/3886	156/3962 (from '91)	156/3962	**Length, in/mm**
57/1448	64/1626	64/1626	**Width, in/mm**
50/1270	52/1290	52/1290	**Height, in/mm**
96/2438	98/2490	98/2490	**W/base, in/mm**
47/1194	50/1280	50/1280	**Front track, in/mm**
49/1245	56/1420	56/1420	**Rear track, in/mm**
about 1870/848	about 1985/920	about 1959/920	**Weight, lb/kg**
2 str, 4 str only	2 str, 4 str only	2 str, 4 str only	**Body styles**
125, 1985-87	357, 1988-92	approx 500 to end '96	**Production**

APPENDIX E Plus 8 technical specifications

	1968	**1973**	**1976**	**1984**	**1990**
Engine	**90-degree 16-valve ohv pushrod V8**				
Bore/stroke, mm	88.90 x 71.12	88.90 x 71.12	88.90 x 71.12	88.90 x 71.12	94.0 x 71.12
Capacity	3528cc (215cu in)	3528cc (215cu in)	3528cc (215cu in)	3528cc (215cu in)	3946cc (241cu in)
Compression	10.5:1	9.35:1	9.25:1	9.75:1	9.35:1
Induction	Twin SU HS6	SU H1 F6	SU/Stromberg	EFI	EFI
Power/rpm	168bhp at 5200	143bhp at 5000	155bhp at 5000	190 at 5250	185 at 4750
Torque/rpm	210lb. ft at 2700	202lb. ft at 2750	202lb. ft at 2750	220lb. ft at 4000	235lb. ft at 2600
Block/head	aluminium/aluminium	aluminium/aluminium	aluminium/aluminium	aluminium/aluminium	aluminium/aluminium
Valvegear	ohv, pushrods operating hydraulic tappets from a single central camshaft				
Transmission	Front engine, rear drive through manual gearbox. One-off auto only				
Clutch	Single-dry-plate	Single-dry-plate	Single-dry-plate	Single-dry-plate	Single-dry-plate
Gearbox	4-speed Moss (484 cars); from April '72, 4-speed Rover (702 cars); from October '76 to date, 5-speed Rover				
Chassis	Z-section side rails with cross-bracing. Ash frame supports hand-formed bodywork, steel to begin with, later steel and/or aluminium				
Steering	Cam Gear cam-and-peg with collapsible column. Jack Knight rack-and-pinion '83, first as option, later as standard				
Front suspension	Independent by lubed sliding stub axle acting on coil springs and hydraulic dampers				
Rear suspension	Live rear axle, longitudinal leaf springs, hydraulic lever-arm dampers. Telescopic dampers from '91				
Brakes	11in disc/9in drum. Lockheed replaced by servo-assisted Girling from '93				
Wheels	Originally, special 15in alloys. From Sept '76, 14in alloys. From '82, new 15in alloys. Optional 7 x 16 centre-lock wires from '91				
Tyres	Originally 185/VR15. To 205/60VR15 (205/55 on 'wires) in '97				
Dimensions					
Length	Originally 152in/3861mm. From '76 156/2962				
Width	Originally 57in/1449mm, then 59/1499 ('73), then 62/1575 ('76), then 63/1600				
Height	Between 48 and 52 inches (1219-1290mm) depending on year, tyres and weight				
Wheelbase	98in/2489mm all models				
Front track	Originally 49in/1245mm, then 51/1295 ('73), then 52/1345 ('76)				
Rear track	Originally 51in/1295mm, then 52/1321 ('73), then 54/1372 ('76)				
Weight	Around 1900lb (862kg) initially, depending on equipment, rising to 2072lb (940kg) for '97 3.9-litre cat-cleaned car				
Prices	£1487 2s 9d as tested by *Motor* magazine in September 1968 with optional front seat belts. Price in January '97, £29,326 before extras. Used Plus 8s (as of Jan '97) from £10,000 to £35,000				
Body styles	All two-seater sports except for one-off four-seater special				
Production	Approximately 4600 made to the end of 1996. Add approximately 200 for each year thereafter				

APPENDIX F

Useful addresses

United Kingdom

DISTRIBUTORS

Central London
Wykehams Ltd
6 Kendrick Place, Reece Mews,
South Kensington, London SW7 3HF
Tel 0171 589 6894 Fax 0171 589 8886

Greater London
F H Douglass
1a South Ealing Road,
London W5 4QT
Tel 0181 567 0570 Fax 0181 840 8132

Libra Motive Ltd
2-10 Carlisle Road, Colindale,
London NW9 0HN
Tel 0181 205 4488 Fax 0181 205 2233

Avon
John Dangerfield /St George
115 Staple Hill Road, Fishponds,
Bristol BS16 5AD
Tel 01179 494747 Fax 01179 497728

Bedfordshire
Allon White & Son Ltd
The Morgan Garage, High Street,
Cranfield, Bedford MK43 0BS
Tel 01234 750205 Fax 01234 751736

Berkshire
Richard Thorne Classic Cars
Bloomfield Hatch, Mortimer,
Reading, Berkshire RG7 3AD
Tel 01734 333633 Fax 01734 333715

SGT
Station Road, Taplow,
Nr Maidenhead SL6 0NT
Tel 01628 605353 Fax 01628 663467

Cheshire
Cheshire Morgans
Wolf Garage, Ashley Road, Hale,
Altrincham, Cheshire WA15 9NQ
Tel 0161 929 1208 Fax 0161 929 9247

Co. Durham
I & J Macdonald Ltd
Howden Works, Lanchester,
Durham DH7 0QR
Tel 01207 520916 Fax 01207 529860

Devon
Phoenix Motors
The Green, Woodbury,
Exeter EX5 1LT
Tel 01395 232255

Essex
Cliffsea Cars
654 Sutton Road, Southend on Sea,
Essex SS2 5PX
Tel 01702 602042 Fax 01702 469188

Kent
Brands Hatch Morgan Ltd
45 Maidstone Road, Borough Green,
Nr Sevenoaks, Kent TN15 8HA
Tel 01732 882017 Fax 01732 886387

Lancashire
Lifes Motors Ltd
West Street, Southport PR8 1QN
Tel 01704 531375 Fax 01704 531126

West Midlands
Mike Duncan at Heart of England Morgans
Hartlebury Garage, Worcester Road,
Hartlebury, Worcestershire DY11 7XH
Tel 01299 250025 Fax 01299 250012

Wiltshire
Burlen Services
Spitfire House, Castle Road,
Salisbury SP1 3SA
Tel 01722 412500 Fax 01722 334221

Yorkshire
Otley Motors Ltd
Cross Green, Pool Road, Otley LS21 1HE
Tel 01943 465222 Fax 01943 850632

Scotland
Parker of Stepps
Hayston Garage, 38 Glasgow Road,
Kirkintilloch, Glasgow G66 1BJ
Tel 0141 776 1708 Fax 0141 775 0255

Thomson & Potter Ltd
High Street, Burrelton,
Blairgowrie, Perthshire PH13 9NX
Tel 01828 670247 Fax 01828 670248

SERVICE FACILITIES

Autorapide Ltd
The Wells Road, Latcham,
Wedmore, Somerset BS28 4SB
Tel 01934 713723 Fax 01934 713266

Harpers
Essex Lane, Hunton Bridge,
Herts WD4 8PN
Tel 01923 260299 Fax 01923 264813

Melvyn Rutter Ltd
The Morgan Garage, Little Hallingbury,
Nr Bishop's Stortford, Herts CM22 7RA
Tel 01279 725725 Fax (Parts) 01279 600498
Fax (Sales & Service) 01279 726901

Perranwell Garage
Perranwell Station, Truro TR3 7PT
Tel & Fax 01872 863037

Techniques
Porter Bros Garage, Station Road,
Radlett, Herts WD7 8JX
Tel 01923 853600 Fax 01923 853343

Worldwide

DISTRIBUTORS

Australia
Morgan Sports Car Distributors Australia
8 Parker Street, Castlemaine, Victoria 3450
Tel 018 509780 Tel/Fax 0354 722025

Austria
Hammerschmid GesmbH
Lüszstrasze 2, 2521 Trumau, Austria
Tel 0043 2253 6666 Fax 0043 2253 8288

Belgium
Garage Albert Anc Ets Stammet Et Fils SPRL
84-86 Rue Osseghem, 1080 Brussels
Tel 0032 2410 6443 Fax 0032 2410 8965

Canada
CMC Enterprises (1990) Inc
12944 Albion Vaughan Rd, RR3,
Bolton, Ontario L7E 5R9
Tel 001 905 857 3210 Fax 001 905 857 3210

Cyprus
Reliable Motor Cars Ltd
Corner Grivas Dhigenis & Prodromou
P O Box 5428, 1309 Nicosia, Cyprus
Tel 00357 2 360903 Fax 00357 2 360904/472768

Denmark
Alan Hall, Dansk Morgan Agentur
Hjelmslevvej 2, Hemstok, 8660 Skanderborg
Tel 0045 8651 1552 Fax 0045 8651 1559

France
Jacques Savoye SA
237 Boulevard Pereire, 75017 Paris
Tel 0033 01 457 48280 Fax 0033 01 457 26546

Germany
Merz & Pabst
Alexanderstrasse 46, 70182 Stuttgart 1
Tel 0049 7112 33111 Fax 0049 7112 33638

Flaving
Hochstrasse 4, 59425 Unna
Tel 0049 2303 251910 Fax 0049 2303 22376

Greece
Elmec Sport S.A.
96 Voulliagmenis Ave, & 65 Miltiadou Str,
16674 Glyfada, Athens, Greece
Tel 0030 1 964 8337 Fax 0030 1 964 8335/6

Holland
B V Nimag
Reedijk 9, P O Box 3250, 3274 ZH Heinenoord
Tel 0031 186 607707 Fax 0031 186 607980

Ireland
Scott MacMillan
Holybrooke Hall, Bray,
Co. Wicklow, Ireland
Tel 00353 1286 0382 Fax 00353 1286 3039

Italy
Armando Anselmo
Via Vincenzo Tiberio 64, 00191 Rome
Tel 00396 3332307

Japan
Morgan Auto Takano Ltd
9-25 2 Chome, Tsumada-Minami, Atsugi
Shi, Kanagawa Ken 243
Tel 0081 337586721 Fax 0081 337586761

Luxembourg
Yesteryear Luxembourg Import Ltd S.A.
19 Rue du Parc, L-8083 Bertrange
Tel 00352 504479 Fax 00352 502939

New Zealand
Auto Restorations Ltd
P O Box 22273, 148 Carlyle Street,
Christchurch
Tel 0064 3366 9988 Fax 0064 3366 5079

Norway
Hallan AS
Roadster Square, Malmoegt 7,
0566 Oslo
Tel 0047 223 78801 Fax 0047 223 84154

Portugal
Manuel F Monteiro & Filho
Quinta do Paizinho,
Rua do Proletariado,
Edificio Monteiros,
2795 Linda-a-Velha
Tel 00351 1417 0514 Fax 00351 1417 1914

Spain
Tayre SA
Principe de Vergara 253,
28016 Madrid
Tel 0034 1457 7634 Fax 0034 1457 7633

South Africa
Terry M Allan
P O Box 71288, Bryanston,
2021 Rep. of S. Africa
Tel 0027 11811 2690 Fax 0027 11811 1160

Sweden
Wendels Bilförsäljning AB
Hornyxegatan 12, Box 9021,
200 39 Malmo
Tel 0046 4021 8000 Fax 0046 4021 0015

Switzerland
Garage de l'Autoroute
Centre commercial de Signy,
1274 Signy S/Nyon
Tel 0041 22361 0931/32 Fax 0041 22361 1385

Rolf Wehrlin
Haupstrasse 132, Aesch BL

USA
Isis Imports Ltd
P O Box 2290, Gateway Station
San Francisco, CA 94126
Tel 001 415 433 1344 Fax 001 415 788 1850

Cantab Motors Ltd
Valley Industrial Park,
12 E Richardson Lane,
Purcellville, VA 20132
Tel 001 540 338 2211 Fax 001 540 338 2944